Impact of COVID-19 Pandemic on Global Diseases and Human Well-Being

Impact of COVID-19 Pandemic on Global Diseases and Human Well-Being

Editor

Giorgio I. Russo

MDPI • Basel • Beijing • Wuhan • Barcelona • Belgrade • Manchester • Tokyo • Cluj • Tianjin

Editor
Giorgio I. Russo
Urology
University of Catania
Catania
Italy

Editorial Office
MDPI
St. Alban-Anlage 66
4052 Basel, Switzerland

This is a reprint of articles from the Special Issue published online in the open access journal *Journal of Clinical Medicine* (ISSN 2077-0383) (available at: www.mdpi.com/journal/jcm/special_issues/impact_pandemic).

For citation purposes, cite each article independently as indicated on the article page online and as indicated below:

LastName, A.A.; LastName, B.B.; LastName, C.C. Article Title. *Journal Name* **Year**, *Volume Number*, Page Range.

ISBN 978-3-0365-5252-1 (Hbk)
ISBN 978-3-0365-5251-4 (PDF)

Contents

About the Editor

Giorgio I. Russo

Prof. Giorgio I. Russo has been an Associate Professor in Urology at the University of Catania since 2020. He has been a research and Clinical fellow at the Department of Urology at the University of Tübingen with Prof. A. Stenzl from July 2017 to January 2018.

He is the Chairman of the EAU Young Academic Urologists –Men's Health working group and Ex-Officio member of the YOU and the ESAU section. He has helped with assisted reproduction techniques at the U.M.R. HERA in Catania.

He is a statistical analyst at the Urological clinic of the University of Catania, with more than 200 public publications (https://www.ncbi.nlm.nih.gov/pubmed/?term=russo+gi).

In 2019, he was awarded with the Ferring 2019 Grant for a project on bladder cancer and 2021 with the Young Academic Urologist Award of the European Association of Urology.

He has been a public speaker at many international (EAU, AUA, SIU) and national congresses in the field of andrology, sexual medicine and uro-oncology.

Journal of
Clinical Medicine

Editorial

Impact of the COVID-19 Pandemic on Global Diseases and Human Well-Being

Arturo Lo Giudice, Maria Giovanna Asmundo, Sebastiano Cimino and Giorgio Ivan Russo *

Department of Surgery, Urology Section, University of Catania, 95124 Catania, Italy;
arturologiudice@gmail.com (A.L.G.); mariagiovannaasmundo@gmail.com (M.G.A.);
ciminonello@hotmail.com (S.C.)
* Correspondence: giorgioivan.russo@unict.it

1. Introduction

This editorial of the Special Issue "Impact of SARS-CoV-2 Pandemic on Global Diseases and Human Well-Being" aims to portray the repercussions of the novel COVID-19 emergency on a wide range of health issues.

The novel acute respiratory syndrome caused by COVID-19 quickly spread after its very first detection on 31 December 2019 in Wuhan, China [1]. Due to the pandemic emergency, many social restrictions were applied and health systems took unprecedented stringent measures that unavoidably influenced people's lives and disease management. However, according to the study conducted by Kim et al., the incidence of COVID-19 infection is different among various income groups; in detail, these authors analyzed low-, middle- and high-income populations to verify if any increased prevalence of COVID-19 exists in these populations. In this study, an increased possibility of viral exposure was detected among low-income populations, probably due to their living and working environmental conditions, such as poor hygiene, less access to healthcare and crowded living conditions [2]. In conclusion, differences in mortality are reported for people of different income levels in Korea [3].

Since the coronavirus pandemic has undoubtedly impacted every person's life, it is easy to understand the great effort made to realize a new vaccine. Even though vaccination played a key role in the current emergency scenario, it also poses several problems, including the possibility of side effects that lead to a diffuse rejection of the vaccine by patients [4]. The systematic review conducted by Sessa et al. clarifies that, even though the total rate of severe side effects related to COVID-19 vaccines is very low, it is important to report them in order to advance our knowledge and support our decisions. According to this study and considering the extremely small number of subjects involved in these rare adverse effects (3 to 10 cases per million), it is possible that the thrombotic thrombocytopenia caused by the COVID-19 vaccine may be multifactorial or deeply influenced by genotype; otherwise, several hypotheses exist: it may be caused by the possible cross reactivity of antibodies against the SARS-CoV-2 spike protein with PF4, interactions between spike protein and platelets, the platelet expression of adenoviral proteins and the resulting immune response [5,6].

2. Chronic Disease

The COVID-19 pandemic has affected many people worldwide, with serious consequences for many patients. When dealing with the high spread of the novel acute respiratory syndrome caused by coronavirus-19, the most vulnerable patients were considered the most important. Moreover, most patients suffering from chronic diseases such as hypertension, diabetes, chronic kidney disease, and hypercholesterolemia were classified into the high-risk category. In the study conducted by Scicali et al., the impact of the direct and indirect effects of SARS-CoV-2 infection in subjects with familial hypercholesterolemia (FH) is evaluated. Predictably, the percentage of patients affected by FH

Citation: Lo Giudice, A.; Asmundo, M.G.; Cimino, S.; Russo, G.I. Impact of the COVID-19 Pandemic on Global Diseases and Human Well-Being. *J. Clin. Med.* **2022**, *11*, 4489. https://doi.org/10.3390/jcm11154489

Received: 26 July 2022
Accepted: 29 July 2022
Published: 1 August 2022

Publisher's Note: MDPI stays neutral with regard to jurisdictional claims in published maps and institutional affiliations.

who consulted lipidologists and/or cardiologists and/or subdued vascular imaging was lower after lockdown compared to the period before, especially because of the fear of contagion. Finally, according to the cohort of 260 patients who took part in the study, the percentage of subjects affected by SARS-CoV-2 was 7.3% and none of them required hospital assistance. Moreover, this study evidenced that the percentage of lipids, through lipid profile evaluation, was lower after lockdown than before (56.5% vs. 100.0%, $p < 0.01$), with a reduction in HDL-C (47.78 ± 10.12 vs. 53.2 ± 10.38 mg/dL, $p < 0.05$), and a relevant increase in non-HDL-C (117.24 ± 18.83 vs. 133.09 ± 19.01 mg/dL, $p < 0.05$). This finding may be explained by the unregulated and sedentary lifestyle that characterized the pandemic period [7]. Moreover, in the pandemic scenario, there has been great interest in the association between SARS-CoV-2 and kidney function; in fact, since the very beginning of the pandemic, several studies analyzed the impact of COVID-19 from different points of view. Precisely, the systematic review and meta-analysis conducted by Chen et al. evaluated the mortality rate, intensive care unit admission, invasive mechanical ventilation, acute kidney injury, kidney replacement therapy and graft loss in the adult kidney transplant population with COVID-19. As is easy to imagine, kidney transplant patients, especially due to their immunocompromised systems, are continually exposed to complications such as opportunistic infections or lymphoproliferative diseases [8]. The higher predisposition and diminished response to infection in the adult kidney transplant population with SARS-CoV-2 disease results in a higher percentage of mortality compared to the general population. In fact, the authors demonstrated increased rates of adverse outcomes among transplanted patients: mortality—21%; admission to intensive care units—26%; intensive mechanical ventilation among those who required admission in intensive care units—72%; acute kidney injury—44%; kidney replacement therapy—12%; and graft loss—8%. Moreover, a higher risk of mortality for elder patients has been registered too [9].

Since the previously mentioned higher risk of mortality and adverse outcomes in patients with chronic diseases is a pressing issue, it is fundamental to assess patients' increased risk when attending hemodialysis treatments, peritoneal dialysis follow-up or after-transplant visits. In this scenario, the prevalence of SARS-CoV-2 infection among the general population plays a fundamental role in the assessment of the augmented risk of COVID-19 infection among chronic disease patients. As reported in a meta-analysis including 1389 patients, COVID-19 seems to augment the possibility of suffering major consequences among frailer populations. In fact, clinical manifestations of COVID-19 infection are reported to be more serious in aged and pluri-pathological patients. As well as hypertension, diabetes, chronic obstructive pulmonary disease and chronic heart disease, underlying kidney disease seems to be related to a higher incidence of mortality and complications too [10]. In this context, being aware of prevalence and screening test precision will aid doctors in the application of preventative measures to limit COVID-19 spread among more vulnerable populations—such as CKD patients. Unluckily, several factors, such as the sensibility and specificity of screening tests, the type of samples, and the timing of the screening, may alter the final results. However, it is undoubted that to reduce the risks of spread between patients and suffering major consequences, clinicians have to carefully manage the results of the screening test for SARS-CoV-2 [11]. Considering the ever-increasing necessity of an early detection of COVID-19 infection among chronic disease patients, Vial et al. investigated the application of tools routinely used to monitor hemodialysis patients as detection indicators for SARS-CoV-2 disease. In detail, based on a low-cost triage tool, the authors observed that total leukocytes were appreciably lower in patients affected by COVID-19 (4.1 vs. 7.4 G/L, $p = 0.0072$) and were characterized by lower levels of eosinophils (0.01 vs. 0.15 G/L, $p = 0.0003$) and neutrophils (2.7 vs. 5.1 G/L, $p = 0.021$). Moreover, eosinophil count below a certain range (0.045 G/l) seems to be indicative of COVID-19 infection with an AUC of 0.9 [95% CI 0.81–1] ($p < 0.0001$), sensitivity of 82%, specificity of 86%, a positive predictive value of 82%, a negative predictive value of 86% and a likelihood ratio of 6.04. In conclusion, these results suggest the possibility of the early detection ofSARS-CoV-2 by a cheap and easily accessible tool such as CBC [12].

3. Everyday Life

The COVID-19 pandemic has affected modern society both from a strictly health perspective and from a social perspective, regarding everyday life implications. Of course, some people in particular situations have been affected more than others, and our editorial is focused on the stress perceived by the caregivers of patients with Alzheimer's disease (AD) during the pandemic.

The "caregiver burden" consists of the emotional, physical, social, or financial burden that the caregiver feels in caring for his/her family member. The caregivers' perceptions of stress can be influenced by psychosocial factors, such as kinship, cultural and social aspects [13,14].

A study has been published in our editorial that evaluates the psychological responses of caregivers of individuals with dementia during the COVID-19 pandemic lockdown; a cross-sectional survey using an anonymous online questionnaire was used [15].

The questionnaire included three sections that presented closed-ended questions with five-point Likert scales and binary-type questions (except for the first one, which collected socio-demographic data). This survey consisted of (1) caregivers' sociodemographic data (gender, age, education, residential position in the last 14 days, marital status, working status, and type of relationship with the patient being assisted) and information about the patient's illness; (2) psychological scales to assess the impact of the COVID-19 pandemic; and (3) tools investigating caregivers' physical and mental health.

Eighty-four AD patients' caregivers were involved in this study by answering an online questionnaire. The data showed that caregivers were affected by high burden and stress; in fact, they obtained a high mean score on the Perceived Stress Scale. Moreover, caregivers' burden was mainly related to their patients' physical difficulties (assessed by Caregiver Burden Inventory—Physical Burden) and perception of losing time (assessed by Caregiver Burden Inventory—Time-dependence Burden). Moreover, caregivers perceived their quality of life as very low (assessed by Short Form-12 Health Survey Physical and Short Form-12 Health Survey Mental Health). Finally, this study demonstrated that participants mostly used dysfunctional coping strategies, such as avoidance strategies (assessed by Coping Orientation to Problem Experiences—Avoidance Strategies); however, these approaches did not affect their stress levels.

4. Mental Health

The deterioration of sociopsychological status and mental health due to governmental restrictions after the spread of COVID-19 has been widely investigated in our Special Issue, with one study aiming to clarify who, when, how, and why individuals died by suicide in Japan during the COVID-19 pandemic. This study assessed a change in the percentage of suicide during the pandemic compared to the pre-pandemic period using a governmental database that divided subjects by prefectures, gender, age, means, motive, and household factors using a linear mixed-effects model [16].

Suicide mortality decreased during the first stay-home order and increased after the first stay-home order ended. Furthermore, the direct health hazard of COVID-19 itself functioned as a suicide suppressor; nevertheless, the protraction of the COVID-19 pandemic period deeply contributed to the increasing incidence of suicide, especially for females. Contrary to nationwide fluctuation patterns, the suicide mortality incidence in metropolitan regions for both genders, male and female, did not decrease during the first stay-home order. Other factors, such as gender (female), age (adolescents), one-person household residents, and living in metropolitan areas, were possible risks of increasing suicide mortality in 2020. The reduction in SMR-S in all 47 prefectures during the first stay-home order might be compared to the "honeymoon period" phenomenon. The stabilization of suicide mortality observed during each stay-home order may also suggest people becoming accustomed to the pandemic.

5. Sexual Health

Pre-COVID

An Italian study carried out between 1 June and 31 December 2019, involving people of any gender and sexual orientation, aimed to describe the most common kinds of contemporary sexual behaviors in Italy prior to the COVID-19 pandemic. Participants were recruited via social media posts on Facebook and Instagram, and Google Forms was used to create and deliver the survey online. Each of the 12,590 people who took part in this study consented to fill out the survey. The survey questions assessed a range of factors, including the frequency and pleasure of various sexual activities (self-stimulation, being masturbated by a partner, masturbating one's partner, receiving and providing oral sex, vaginal penetration, receiving and providing anal penetration), sexual satisfaction, the frequency of orgasms, triggers for auto-eroticism, the use of sex toys, the pleasantness of various sexual fantasies, pornography use, betrayal, traumatic sexual experiences, stress, contraception, protection against sexually transmitted infections, the use of medications or drugs, the use of dating apps or sites and sexting. Most participants were heterosexual, 10,153 (80.6%), followed by homosexuals (234), bisexuals (2087; 16.6%), and pansexuals (83; 0.7%). Only 20–30% of participants in the poll used sex toys, while the majority watched pornography on a weekly basis (27.8 %) and alone (80%). Having intercourse in public, having sex with multiple people at once, having sex while blindfolded, being tied up, and watching a naked person are the fantasies that most stimulate and excite the participants. About 80–90% of the respondents indicated that they did not engage in anal intercourse; it is probable that, in Italy, sexual independence and the urge to test out novel sexual practices may be eclipsed by a widespread sense of shame [17].

It was shown that crises can alienate loved ones; moreover, the loss of a home can alter daily habits and couples' sex lives. The repercussions of the lockdown on the Italian population's sex lives are less known. In Italy, crises that have changed peoples' habits (e.g., the L'Aquila earthquake) have been widely studied. After this earthquake, a high rate of sexual dysfunction-related symptoms was reported among young adults, particularly in subjects experiencing post-traumatic stress disorder [18]. The lockdown period has likely led to changes in Italians' sex lives. Precisely, in this Special Issue, we aimed to investigate if any change in adult men and women's sexual behaviors occurred during lockdown.

An Italian, multicenter, cross-sectional study, was conducted in 15 urological centers. This research was performed through a Google Forms online survey, from 4 May 2020 (50 days after the start of the lockdown) to 18 May 2020. Inclusion criteria were sexually active subjects in stable relationships for at least 6 months; any age and gender were included. Exclusion criteria were subjects who were affected by COVID-19, single or sexually inactive. In the end, 2149 participants were enrolled in this study. The results showed that 29% of subjects considered that their sex lives with their partners had "much or very much" deteriorated during the lockdown period; otherwise, 49% considered it to be "much or very much" improved during the same period. Finally, 225 did not report any deterioration or improvement.

Among participants who reported an improvement in their sex lives with their partners, the greatest percentage was represented by women; this result was found to be significantly associated with cohabitation, having a stable relationship for more than 5 years and being married without children. No patients of any gender reported having sexual dysfunction. On the other hand, most of the participants who reported a worsening of their sex lives with their partners did not live with their partners during lockdown (73.4%). Among cohabitees, most had sons (82%) and a stable relationship for more than 5 years (81.7%). Among the women that reported a worsening of their sex lives with their partners, there were no sexual disfunction but a higher level of anxiety, tension, fear, and insomnia; on the other hand, men who reported worsening of their sex lives with their partners had a higher rate of mild erectile dysfunctions, orgasmic dysfunctions, and low sexual satisfaction within the previous 4 weeks.

However, despite the impossibility of meeting friends and relatives during lockdown, a reconciliation took place between cohabiting couples. Most people expressed satisfaction having their partners at home. The improvement was reported primarily in participants who had been in stable relationships for more than 5 years, probably because the increased time spent together favored the rediscovery of a feeling that the couple might have lost in their life routines. Moreover, spending entire days at home can stimulate and facilitate common interests between partners—the sharing of hobbies or daily practices that normally could not be shared because of a lack of time. Participants over the age of 40 improved more than those under the age of 40, probably because most younger participants did not live with their partners during that time [19].

Author Contributions: Conceptualization, A.L.G., M.G.A. and G.I.R.; methodology, A.L.G., M.G.A. and G.I.R.; writing—original draft preparation, A.L.G., M.G.A. and G.I.R.; writing—review and editing, A.L.G., M.G.A. and G.I.R.; supervision, S.C. All authors have read and agreed to the published version of the manuscript.

Funding: This research received no external funding.

Conflicts of Interest: The authors declare no conflict of interest.

References

1. Sharma, A.; Ahmad Farouk, I.; Lal, S.K. COVID-19: A Review on the Novel Coronavirus Disease Evolution, Transmission, Detection, Control and Prevention. *Viruses* **2021**, *13*, 202. [CrossRef] [PubMed]
2. Tsai, J.; Wilson, M. COVID-19: A potential public health problem for homeless populations. *Lancet Public Health* **2020**, *5*, e186–e187. [CrossRef]
3. Kim, S.Y.; Yoo, D.M.; Min, C.; Choi, H.G. The Effects of Income Level on Susceptibility to COVID-19 and COVID-19 Morbidity/Mortality: A Nationwide Cohort Study in South Korea. *J. Clin. Med.* **2021**, *10*, 4733. [CrossRef] [PubMed]
4. Scavone, C.; Sessa, M.; Clementi, E.; Rossi, F.; Capuano, A. Italian Immunization Goals: A Political or Scientific Heated Debate? *Front. Pharmacol.* **2018**, *9*, 574. [CrossRef] [PubMed]
5. Sessa, F.; Salerno, M.; Esposito, M.; Di Nunno, N.; Zamboni, P.; Pomara, C. Autopsy Findings and Causality Relationship between Death and COVID-19 Vaccination: A Systematic Review. *J. Clin. Med.* **2021**, *10*, 5876. [CrossRef] [PubMed]
6. Rzymski, P.; Perek, B.; Flisiak, R. Thrombotic Thrombocytopenia after COVID-19 Vaccination: In Search of the Underlying Mechanism. *Vaccines* **2021**, *9*, 559. [CrossRef] [PubMed]
7. Scicali, R.; Piro, S.; Ferrara, V.; Di Mauro, S.; Filippello, A.; Scamporrino, A.; Romano, M.; Purrello, F.; Di Pino, A. Direct and Indirect Effects of SARS-CoV-2 Pandemic in Subjects with Familial Hypercholesterolemia: A Single Lipid-Center Real-World Evaluation. *J. Clin. Med.* **2021**, *10*, 4363. [CrossRef] [PubMed]
8. Yeo, W.-S.; Ng, Q.X. Biomarkers of immune tolerance in kidney transplantation: An overview. *Pediatric Nephrol.* **2022**, *37*, 489–498. [CrossRef] [PubMed]
9. Chen, J.-J.; Kuo, G.; Lee, T.H.; Yang, H.-Y.; Wu, H.H.; Tu, K.-H.; Tian, Y.-C. Incidence of Mortality, Acute Kidney Injury and Graft Loss in Adult Kidney Transplant Recipients with Coronavirus Disease 2019: Systematic Review and Meta-Analysis. *J. Clin. Med.* **2021**, *10*, 5162. [CrossRef] [PubMed]
10. Henry, B.M.; Lippi, G. Chronic kidney disease is associated with severe coronavirus disease 2019 (COVID-19) infection. *Int. Urol. Nephrol.* **2020**, *52*, 1193–1194. [CrossRef] [PubMed]
11. Martino, F.; Amici, G.; Grandesso, S.; Ferraro Mortellaro, R.; Lo Cicero, A.; Novara, G. Analysis of the Clinical and Epidemiological Meaning of Screening Test for SARS-CoV-2: Considerations in the Chronic Kidney Disease Patients during the COVID-19 Pandemic. *J. Clin. Med.* **2021**, *10*, 1139. [CrossRef] [PubMed]
12. Vial, R.; Gully, M.; Bobot, M.; Scarfoglière, V.; Brunet, P.; Bouchouareb, D.; Duval, A.; Zino, H.-O.; Faraut, J.; Jehel, O.; et al. Triage of Patients Suspected of COVID-19 in Chronic Hemodialysis: Eosinophil Count Differentiates Low and High Suspicion of COVID-19. *J. Clin. Med.* **2020**, *10*, 4. [CrossRef] [PubMed]
13. Etters, L.; Goodall, D.; Harrison, B.E. Caregiver burden among dementia patient caregivers: A review of the literature. *J. Am. Acad. Nurse Pract.* **2008**, *20*, 423–428. [CrossRef] [PubMed]
14. Chiao, C.-Y.; Wu, H.-S.; Hsiao, C.-Y. Caregiver burden for informal caregivers of patients with dementia: A systematic review. *Int. Nurs. Rev.* **2015**, *62*, 340–350. [CrossRef] [PubMed]
15. Maggio, M.G.; La Rosa, G.; Calatozzo, P.; Andaloro, A.; Foti Cuzzola, M.; Cannavò, A.; Militi, D.; Manuli, A.; Oddo, V.; Pioggia, G.; et al. How COVID-19 Has Affected Caregivers' Burden of Patients with Dementia: An Exploratory Study Focusing on Coping Strategies and Quality of Life during the Lockdown. *J. Clin. Med.* **2021**, *10*, 5953. [CrossRef] [PubMed]
16. Matsumoto, R.; Motomura, E.; Fukuyama, K.; Shiroyama, T.; Okada, M. Determining What Changed Japanese Suicide Mortality in 2020 Using Governmental Database. *J. Clin. Med.* **2021**, *10*, 5199. [CrossRef] [PubMed]

17. Di Mauro, M.; Russo, G.I.; Polloni, G.; Tonioni, C.; Giunti, D.; Cito, G.; Giammusso, B.; Morelli, G.; Masieri, L.; Cocci, A. Sexual Behaviour and Fantasies in a Group of Young Italian Cohort. *J. Clin. Med.* **2021**, *10*, 4327. [CrossRef] [PubMed]
18. Carmassi, C.; Dell'Oste, V.; Pedrinelli, V.; Barberi, F.M.; Rossi, R.; Bertelloni, C.A.; Dell'Osso, L. Is Sexual Dysfunction in Young Adult Survivors to the L'Aquila Earthquake Related to Post-traumatic Stress Disorder? A Gender Perspective. *J. Sex Med.* **2020**, *17*, 1770–1778. [CrossRef] [PubMed]
19. Costantini, E.; Trama, F.; Villari, D.; Maruccia, S.; Li Marzi, V.; Natale, F.; Balzarro, M.; Mancini, V.; Balsamo, R.; Marson, F.; et al. The Impact of Lockdown on Couples' Sex Lives. *J. Clin. Med.* **2021**, *10*, 1414. [CrossRef] [PubMed]

Journal of
Clinical Medicine

Article

How COVID-19 Has Affected Caregivers' Burden of Patients with Dementia: An Exploratory Study Focusing on Coping Strategies and Quality of Life during the Lockdown

Maria Grazia Maggio [1], Gianluca La Rosa [2], Patrizia Calatozzo [3], Adriana Andaloro [3], Marilena Foti Cuzzola [3], Antonino Cannavò [2], David Militi [4], Alfredo Manuli [2], Valentina Oddo [5], Giovanni Pioggia [6] and Rocco Salvatore Calabrò [7,*]

[1] Department of Biomedical and Biotechnological Science, The University of Catania, 95123 Catania, Italy; mariagraziamay@gmail.com

[2] AOU Policlinico Gaetano Martino, 98125 Messina, Italy; Gialucalarosa05@gmail.com (G.L.R.); antonino.cannavo.85@gmail.com (A.C.); manulialfredo@gmail.com (A.M.)

[3] Studio di Psicoterapia Relazionale e Riabilitazione Cognitiva, 98124 Messina, Italy; patrizia.calatozzo@gmail.com (P.C.); ad.andaloro@gmail.com (A.A.); marilenafoti@yahoo.it (M.F.C.)

[4] Odontostomatology and Dental Surgery Study, 98124 Messina, Italy; Davidmiliti@hotmail.it

[5] Università degli Studi di Messina-Piazza Pugliatti, 1, 98122 Messina, Italy; voddo@unime.it

[6] Institute for Biomedical Research and Innovation, National Research Council of Italy (IRIB-CNR), 98164 Messina, Italy; giovanni.pioggia@irib.cnr.it

[7] IRCCS Centro Neurolesi Bonino Pulejo, 98121 Messina, Italy

[*] Correspondence: salbro77@tiscali.it; Tel.: +39-09-0601-2380

Citation: Maggio, M.G.; La Rosa, G.; Calatozzo, P.; Andaloro, A.; Foti Cuzzola, M.; Cannavò, A.; Militi, D.; Manuli, A.; Oddo, V.; Pioggia, G.; et al. How COVID-19 Has Affected Caregivers' Burden of Patients with Dementia: An Exploratory Study Focusing on Coping Strategies and Quality of Life during the Lockdown. *J. Clin. Med.* **2021**, *10*, 5953. https://doi.org/10.3390/jcm10245953

Academic Editor: Giorgio I. Russo

Received: 21 November 2021
Accepted: 16 December 2021
Published: 18 December 2021

Publisher's Note: MDPI stays neutral with regard to jurisdictional claims in published maps and institutional affiliations.

Abstract: COVID-19 has caused a public and international health emergency, leading to isolation and social distancing. These restrictions have had a significant impact on the caregivers of people with dementia, increasing the burden of patient management. The purpose of this study was to investigate the stress perceived by caregivers of patients with Alzheimer's disease (AD) during the pandemic. We used a cross-sectional survey design to evaluate the caregivers' psychological responses and coping strategies. Eighty-four caregivers of patients with a diagnosis of AD were involved in this study by completing an online questionnaire. They presented a high perception of stress (the Perceived Stress Scale mean ± DS: 33.5 ± 4.5), and their high burden in caring was mainly related to physical difficulties (Caregiver Burden Inventory–Physical Burden mean ± DS: 15.0 ± 2.1) and perception of loss of time (Caregiver Burden Inventory–Time-dependence Burden mean ± DS: 16.5 ± 1.4). Moreover, caregivers perceived their quality of life as very low (Short Form-12 Health Survey Physical mean ± DS: 13.5 ± 2.7; Short Form-12 Health Survey Mental Health mean ± DS: 16.4 ± 4.2). Finally, we found that participants mostly used dysfunctional coping strategies, such as avoidance strategies (Coping Orientation to Problem Experiences–Avoidance Strategies mean ± DS: 39.5 ± 7.1), but these strategies did not affect the stress level of caregivers. Given that caregivers present a high burden and stress, innovative tools could be a valuable solution to investigate and support their emotional and behavioral status during difficult periods, such as the COVID-19 pandemic.

Keywords: Alzheimer's disease; burden; caregiver; dementia; quality of life

1. Introduction

On 30 January 2020, the World Health Organization (WHO) declared the COVID-19 pandemic an international public health emergency [1]. A few weeks after the initial outbreak in China, the total number of cases and deaths exceeded disproportionately those of the previous SARS [2–6]. Standard public health measures, including quarantine, social distancing, and community containment, are being used to curb the pandemic of this respiratory disease, and these new measures have changed the dynamics of social relationships, including relationships between doctors and patients, with regard to those with neuropsychiatric symptoms [6].

For these reasons, various authors have highlighted that intervention on people's mental health is necessary given that COVID-19 has profoundly affected psychosocial status worldwide [7–9]. Isolation and social distancing had a significant impact on the caregiver of elderly people affected by chronic diseases, including dementia [7–9].

Briefly, dementia is a syndrome characterized by progressive degeneration of cognitive functions, causing impairment of normal activities and relationships in daily life [10–12]. Families are very important in the "long-term" management of these patients, for both therapeutic compliance and their needs [10]. Because the worsening of cognitive functions can progressively impair the ability to perform simple but essential tasks in daily life, the physical, psychological, and economic impact of dementia on individuals and their families is inevitable [13]. The "caregiver burden" consists of the emotional, physical, social, or financial burden that the caregiver feels in caring for his/her family member. It is a multidimensional concept related to the caregiver's perception of stress while carrying out his/her care activities, and this can be influenced by psychosocial factors, such as kinship and cultural and social aspects, as well as personal characteristics, including sensitivity and vulnerability to stress [14,15]. An adequate network of services to support patients and their families is essential to reduce the burden of caregivers and delay the possible institutionalization of the patient [9]. Indeed, caregivers spend up to 10 h on daycare for the patient and meet all his/her needs, such as feeding, dressing, washing, therapy, and surveillance. A patient's care is often complicated by behavioral problems, such as agitation, physical and verbal aggression, and disappointments. The load of the caregiver may also affect his/her work and the economic dimension, further causing emotional and psychological stress. Finally, the profound changes in the relationship between patients and caregivers may lead to feelings of frustration, despair, and anger. The coronavirus pandemic, with regard to the restrictive measures, could have made these things worse [16,17].

2. Materials and Methods

2.1. Participants and Settings

We used a cross-sectional survey design to evaluate the psychological response of caregivers of individuals with dementia during the COVID-19 pandemic lockdown, using an anonymous online questionnaire. The online survey was administered through the CAWI (Computer Assisted Web Interviewing) method: the invitation to the questionnaire was sent through the technological means offered by smartphones (i.e., WhatsApp, Facebook, Menlo Park, CA, USA) or by email. The questionnaire compilation was carried out by the online survey platform Google. The participants came from the same geographical area, i.e., the province of Messina to avoid cultural biases.

The primary caregiver was defined as the person who lives with the patient in the same home and takes primary responsibility for providing care to the patient at home.

The caregivers list has been made through the generalities and addresses provided by medical doctors (either neurologists or general practitioners) involved in the care of patients with dementia. One hundred fifty individuals were initially contacted by their clinicians, who were previously informed about the research. About 120 of them provided consent to enter the study protocol, but not all of them met the inclusion criteria. To be included in the study, caregivers had to (i) be at least 18 years of age and (ii) be the primary caregiver of a patient affected by AD.

The final sample consisted of 84 primary caregivers of patients with AD (76.2% females; mean age of years ± DS: 45.7 ± 1.3), living in the province of Messina, Italy (Table 1).

Table 1. Descriptive analysis of patients' and caregivers' characteristics.

Patients	84
Age (years)	62.9 ± 4.1
Caregivers	84
Relation to patients	
Son/Daughter	54 (64.3%)
Spouse/Partner	23 (27.3%)
Other	7 (8.4%)
Age (years)	45.7 ± 9.3
Gender	20 (23.8%)
Male	64 (76.2%)
Female	
Education	15.38 ± 2.38
Professions	
Freelancer	17 (21.0%)
Employee	41 (48.0%)
Housewife	16 (19.0%)
Other	10 (12.0%)
Marital Status	
Single	35 (41.7%)
Married	42 (50.0%)
Divorced	7 (8.3%)
Sons	
Yes	45 (53.6%)
No	39 (46.4%)

Mean ± standard deviation was used to describe continuous variables; proportions (numbers and percentages) were used to describe categorical variables.

2.2. Procedures

Following the restrictive measures adopted by the Italian Government to deal with the pandemic, given that it was necessary to minimize face-to-face interactions and stay at home, we asked participants to fill out the online questionnaire.

They completed the questionnaires in Italian through an online survey platform ("Google Form", Google LLC, Mountain View, CA, USA). Data collection took place from 1 April to 20 May 2020, i.e., during the first Italian lockdown.

The study complies with the principles contained in the Helsinki Declaration, and all participants provided informed consent to participate.

2.3. Survey Development

The questionnaire included three areas that collect closed-ended questions with evaluation on 5-point Likert scales and binary type (except for the first one that collected socio-demographic data). The survey consisted of (1) caregivers' sociodemographic data (gender, age, education, residential position in the last 14 days, marital status, working status, type of relationship with the patient being assisted) and information about the patient's illness, (2) psychological scales to assess the impact of the COVID-19 epidemic, and (3) tools investigating caregivers' physical and mental health, i.e., the Perceived Stress Scale [18], the Coping Orientation to Problems Experienced-New Italian Version (COPE-NVI) [19], the Caregiver Burden Inventory (CBI) [20], and the 12-Item Short Form Survey (SF-12) [21] (Table 2).

From the psychometric perspective of scale evaluation, Cronbach's alpha measures internal consistency across the set of individual items. Specifically, they describe the dimension of each clinical tool. In this context, we calculated Cronbach's alpha for each dimension, except for stress level (SSP) because it consists of a single item (i.e., alpha is not available). As shown in Table 2, the items defined for the three dimensions (i.e., COPE-NVI, CBI, and SF-12) are "reliable" in capturing the characteristics of the specific dimension because they exceed the threshold of 0.70.

Table 2. Clinical assessment tools.

Test/Scale Description	Description
PSS	The Perceived Stress Scale (PSS) is the most widely used psychological instrument for measuring the perception of stress. It is a measure of the degree to which situations in one's life are appraised as stressful. Items were designed to tap how unpredictable, uncontrollable, and overloaded respondents find their lives. The scale also includes a number of direct queries about current levels of experienced stress. The items are easy to understand, and the response alternatives are simple to grasp. The questions in the PSS ask about feelings and thoughts during the last month. Regarding the psychometric properties of PSS, it has been shown that it can be used reliably and repeatably to measure perceived stress.
COPE-NVI	The Coping Orientation to Problems Experienced is a self-report questionnaire that considers the coping strategies. The tool consists of five large, essentially independent dimensions: social support, avoidance strategies, positive attitude, problem-solving, and turning to religion. The COPE-NVI can be considered a useful and psychometrically valid tool for measuring coping styles in the Italian context.
CBI	The Caregiver Burden Inventory is a tool for the evaluation of the care load, developed for caregivers of Alzheimer's disease and dementia patients. It is a self-report tool, compiled by the main caregiver. It is a tool for quick completion and easy understanding. Divided into 5 sections, it allows us to evaluate different stress factors: objective load, psychological load, physical load, social load, and emotional load. Regarding the psychometric properties of CBI, it has been shown to be a reliable and repeatable tool.
SF-12	The SF-12 is a self-reported outcome measure assessing the impact of health on an individual's everyday life. It is often used as a quality of life measure. The SF-12 is a shortened version of its predecessor, the SF-36, which itself evolved from the Medical Outcomes Study. The SF-12 was created to reduce the burden of responsibility, and it has been shown that SF-12 can be used reliably and repeatably to measure the quality of life.

2.4. Statistical Analysis

The descriptive statistics were analyzed and expressed as mean $\pm$ standard deviation or as median $\pm$ first/third quartile for continuous variables, as appropriate; frequencies (%) were used for categorical variables. Clinical scale scores were expressed as a mean and standard deviation. The normality of the data was assessed by the Jarque-Bera test: the data met the assumption of normality.

We used linear regressions to calculate the univariate relationship between the perceived level of stress related to the caregiver burden and the scoring of the scales. All tests were two-tailed, with a significance level of $p < 0.05$. Statistical analysis was performed using SPSS Statistic 16.0 (IBM SPSS Statistics, Armonk, NY, USA).

3. Results

Eighty-four participants were included in the study, and all completed the online questionnaire.

As shown in Table 3, caregivers presented a high perception of stress (PSS mean $\pm$ DS: 33.5 $\pm$ 4.5). High levels of physical difficulties (CBI PH mean $\pm$ DS: 15.0 $\pm$ 2.1) and time dependence (CBI TD mean $\pm$ DS: 16.5 $\pm$ 1.4) were frequently present in the caregivers' answers to the questionnaire. The quality of life perceived by caregivers was very low, for the aspects regarding quality of both physical and mental life (SF-12 PH mean $\pm$ DS: 13.5 $\pm$ 2.7, SF-12 MH mean $\pm$ DS: 16.4 $\pm$ 4.2). In addition, we found that participants mostly used dysfunctional coping strategies, such as avoidance strategies (COPE AS mean $\pm$ DS: 39.5 $\pm$ 7.1), with low use of functional strategies, such as orientation to the problem, positive attitude, searching for social support, and transcendent orientation.

Table 3. Average of the clinical scale of caregivers.

Test/Scale	Caregivers	
	Mean ± SD	**Range**
COPE SS	24.2 ± 3.8	14–35
COPE AS	39.5 ± 7.1	19–58
COPE AP	29.2 ± 6.5	14–42
COPE OP	25.3 ± 4.8	14–37
COPE TO	19.4 ± 2.5	13–25
SF-12 PH	13.5 ± 2.7	8–18
SF-12 MH	16.4 ± 4.2	6–27
CBI TD	16.5 ± 1.4	0–20
CBI D	8.2 ± 6.9	0–20
CBI PH	15.0 ± 2.1	0–16
CBI SOCIAL	4.7 ± 5.1	0–19
CBI EMOTIONAL	5.1 ± 3.1	0–16
PSS	33.5 ± 4.5	3–38

Legend: Perceived Stress Scale (PSS) cut-off > 14.0; Coping Orientation to Problem Experiences (COPE) Average (DS) in Italy: Social Support (SS) 27.7(8.4), Avoidance Strategies (AS) 23.5(5.1), Positive Attitude (PA) 30.9(6), Problem Orientation (PO) 32(6.7), Transcendent Orientation (TO) 22.7(5.6); Caregiver Burden Inventory Total (TOT) cut-off > 36.0: Time-dependence Burden (TD), Developmental Burden (D), Physical Burden (PH), Social Burden (Social), Emotional Burden (Emotional); Short Form-12 Health Survey Total (SF-12 TOT) cut-off < 50; Short Form-12 Health Survey Mental Health (SF-12 MH) cut-off < 45.5; Short Form-12 Health Survey Physical (SF-12 Ph) cut-off < 50.

The significant relationship between the perceived level of stress (PSS) and tools investigating caregivers' physical and mental health are reported in Table 4. PSS was not significantly related to any dysfunctional coping strategies; thus, they did not affect the stress level of caregivers. Conversely, PSS had negative and significant relationships with the physical (SF-12 PH) and emotional (SF-12 MH) caregiver quality of life. Specifically, the worse the caregiver's quality of life, the worse the caregiver can manage stress due to their burden, and vice versa. Finally, PSS was positively and significantly related to all the indices of high caregiver burden: time dependence (CBI-TD), development (CBI-D); physical (CBI-PH), social (CBI-SOCIAL), and emotional (CBI-EMOTIONAL). Briefly, the higher the perceived burden of the caregiver, the greater the level of stress they will face.

Table 4. Univariate regression models for a perceived level of stress (PSS).

Variable	Coefficient	*t*-Test	*p*-Value
Constant	16.814	3.2	0.002
COPE SS	0.278	1.3	0.197
Constant	25.334	5.4	0.000
COPE AS	−0.045	−0.38	0.702
Constant	23.747	6.2	0.000
COPE AP	−0.006	−0.05	0.960
Constant	18.727	4.24	0.000
COPE OP	0.191	1.11	0.269
Constant	33.318	5.23	0.000
COPE TO	−0.503	−1.54	0.127
Constant	37.843	9.64	0.000
SF-12 PH	−1.056 **	−3.71	0.000
Constant	44.643	19.46	0.000
SF-12 MH	−1.282 **	−9.49	0.000
Constant	19.207	15.73	0.000
CBI TD	0.509 **	4.5	0.000
Constant	19.069	16.94	0.000
CBI D	0.547 **	5.2	0.000
Constant	18.996	18.43	0.000
CBI PH	0.649 **	6	0.000

Table 4. *Cont.*

Variable	Coefficient	*t*-Test	*p*-Value
Constant	20.433	20.07	0.000
CBI SOCIAL	0.663 **	4.51	0.000
Constant	21.441	22.02	0.000
CBI EMOTIONAL	0.757 **	3.59	0.001

Significance levels of 1% (**) for coefficients by z-test are in bold. Legend: Perceived Stress Scale (PSS) cut-off > 14.0; Coping Orientation to Problem Experiences (COPE) Average (DS) in Italy: Social Support (SS) 27.7(8.4), Avoidance Strategies (AS) 23.5(5.1), Positive Attitude (PA) 30.9(6), Problem Orientation (PO) 32(6.7), Transcendent Orientation (TO) 22.7(5.6); Caregiver Burden Inventory Total (TOT) cut-off > 36.0: Time-dependence Burden (TD), Developmental Burden (D), Physical Burden (PH), Social Burden (Social), Emotional Burden (Emotional); Short Form-12 Health Survey Total (SF-12 TOT) cut-off < 50; Short Form-12 Health Survey Mental Health (SF-12 MH) cut-off < 45.5; Short Form-12 Health Survey Physical (SF-12 Ph) cut-off < 50.

4. Discussion

As people age, there is an increase in the incidence/prevalence of chronic degenerative diseases, such as dementia, i.e., the leading cause of disability at old age [18]. Moreover, medical advances have allowed for an increase in lifespan, even in patients with chronic and disabling diseases. Consequently, the care of patients with chronic disabilities affects the quality of life of caregivers and leads to high stress with important psychosocial problems, especially during pandemics like COVID-19 [19].

The new SARS-CoV-2 pandemic and the consequent limitations have resulted in a significant deterioration in the performance of regular daily living activities, with negative effects on caregivers of patients with dementia, as observed in our sample.

The aim of this study was to investigate the stress and perceived burden of caregivers caring for patients with AD during the COVID-19 pandemic. Using univariate regression analysis, we found that participants with higher levels of perceived stress have their health severely affected. In other words, the personal health condition (both mental and physical) greatly affects the level of stress (as the health condition of the caregiver worsens, the ability to manage stress decreases). At the same time, a higher level of caregiver burden (valid for the five types of CBI explored in this study) can significantly influence the perceived stress level. Additionally, we noted that none of the dysfunctional coping strategies were able to influence the caregiver's perceived stress level, so these strategies were not effective in this COVID-19 framework. Indeed, COVID-19 has profoundly affected the psychosocial state around the world. At an individual level, people experience fear of getting sick or dying with feelings of helplessness for both themselves and their family members [5]. Social restrictions have significantly affected the management of clinics with cancellations or postponements of outpatient visits or rehabilitation activities [7]. Considering the risk of serious COVID-19-related outcomes, most patients with dementia have been forced to stay at home. Hence, the restrictive measures may have worsened the status of patients with dementia, inducing greater discomfort and burden on the caregiver [6,7].

Some authors have shown a worsening of the neuropsychiatric symptoms of patients with dementia, such as anxiety, depression, agitation, and apathy during the COVID-19 pandemic [20,21]. In the presence of psychological and behavioral symptoms, dementia becomes more difficult and stressful to manage than other chronic conditions affecting the elderly. As a consequence, the caregivers have higher emotional and behavioral distress levels [22]. Indeed, caring for people with dementia is very challenging, and family caregivers are at higher risk for physical and mental health problems. This could be due not only to the problems related to the patient's daily care but also to the awareness of the inexorable and uncontrollable progression of the disease [23,24]. Moreover, some studies showed that caregivers are at a greater risk of cardiovascular diseases, such as hypertension, due to the stress-related chronic inflammatory response and excessive sympathetic activation [25].

Concerning the socio-demographic data, our study has highlighted a high level of stress in caregivers, especially in women, married, employed, and, above all, in cases where the caregiver was the patient's son. In particular, stress was perceived as a consequence of the daily needs of patients with AD (and we enrolled only caregivers of this type of dementia), such as assistance in feeding, dressing, bathing, and administering daily therapy. However, stress was higher when caregivers had to deal with neuropsychiatric disorders, such as behavioral problems, agitation, and verbal aggression. According to previous studies, perceived stress primarily affects the perception of time-wasting and physical health, as well as the quality of life. This latter was rated as very low by our sample [26–30].

It is noteworthy that the majority of the sample reported a worsening of stress and family care-related burden during this period, with regard to both clinical and socio-economic aspects [31,32]. In more detail, the reorganization of the healthcare system with the increase of acute wards/services to face COVID-19 and a reduction/closure of social and healthcare services for chronic illness has caused a decrease and/or interruption of the outpatient clinic and/or homecare dedicated to dementia [4]. This has caused an overload on the burden of caregivers who also had to deal with some clinical/health practices for which they did not feel properly prepared or trained [9,32–35]. Furthermore, the reduction of physical contact and social relationships did not allow caregivers to perceive adequate psychophysical and mental support, with a reduction in playful activities, increasing the PSS and worsening their quality of life [9,31–35]. According to recent studies, these sudden changes had an immediate impact on the caregiver's burden by increasing the possibility of precipitating feelings of loneliness, social isolation, and increasing stress levels due to social distancing efforts [9,31–35].

5. Strengths and Limitations

The use of new technologies allowed us to administer the survey. This means of assessment is particularly useful in periods during which social distance is needed to avoid contagions, like during this terrible pandemic. As technological interventions have proven useful in the care of patients with dementia [36,37], future studies could deepen the use of telemedicine for caregivers of patients with AD as an assessing tool and psycho-emotional support for both patients and their caregivers.

The present study had some limitations. The study involved a small sample of caregivers of patients with AD, so there may be difficulties to generalize the results to the patients' population. However, we have focused only on a specific type of dementia, so that findings by our sample might be more homogeneous, given that the different kinds of dementia often have different symptoms and disease progression.

Additionally, this study considers the self-selection issue [38]. The caregivers have voluntarily decided to participate in the questionnaire, probably due to their abilities in using technological devices. Therefore, this selection bias might have affected the accuracy of results, also due to the lack of information concerning the caregivers who were not able to fill out the online questionnaire.

Furthermore, there is no follow-up period, and it is not certain if the results obtained would have lasted over time, also considering the lack of data regarding the burden of caregivers in the pre-COVID era. Future studies are needed to compare the situation resulting from COVID-19 with others occurring out of this health emergency.

We did not collect data on the cognitive, psychological, and physical state of the AD patients, having collected the information from their caregivers, so we can only assume the presence of patient's behavioral changes. Finally, we did not collect data on the amount of time spent by the caregivers with the patients: in the survey, the caregiver was asked to answer only if he/she was the main person in charge of the patient's care, i.e., it was the person who spent more time with the patient than other family members. In future research, it will be necessary to extend the study to a larger sample and increase the involvement of family members and use specific assessment tools for patients as well.

6. Conclusions

To summarize, this study has evaluated the burden of caregivers of patients with AD during the first Italian COVID-19 lockdown. We found that there was an increase in the caregiver's PSS with a worsening of their quality of life. We believe that innovative tools, such as online questionnaires or telemedicine, could be a valuable solution to investigate these concerns and support caregivers of people with dementia during more difficult periods, as the COVID-19 pandemic is. These aspects are fundamental to favor the correct management of chronic diseases at old age. Therefore, healthcare policies and assistance services that provide support to the crucial needs of both frail people and family members caring for them should be developed and promoted.

Author Contributions: Conceptualization, M.G.M. and M.F.C.; methodology, M.G.M., P.C. and A.A.; software, A.C.; validation, A.M., A.C. and G.L.R.; formal analysis, V.O.; investigation, M.G.M., P.C., A.A., A.M. and M.F.C.; resources, P.C.; data curation, A.C. and M.G.M.; writing–original draft preparation, M.G.M., G.L.R. and P.C.; writing–review and editing, R.S.C., P.C. and A.A.; visualization, A.M., G.P. and D.M.; supervision, D.M., G.P. and R.S.C. All authors have read and agreed to the published version of the manuscript.

Funding: This research received no external funding.

Institutional Review Board Statement: The study was conducted according to the guidelines of the Declaration of Helsinki and approved by the Local Institutional Review Board (UNIME 141/20/P).

Informed Consent Statement: Informed consent was obtained from all subjects involved in the study.

Data Availability Statement: The data are available from the online database or on request from the corresponding author.

Conflicts of Interest: The authors declare no conflict of interest.

References

1. Mahase, E. China coronavirus: WHO declares international emergency as death toll exceeds 200. *BMJ* **2020**, *368*, m408. [CrossRef]
2. Hawryluck, L.; Gold, W.L.; Robinson, S.; Pogorski, S.; Galea, S.; Styra, R. SARS Control and Psychological Effects of Quarantine, Toronto, Canada. *Emerg. Infect. Dis.* **2004**, *10*, 1206–1212. [CrossRef]
3. Chen, N.; Zhou, M.; Dong, X.; Qu, J.; Gong, F.; Han, Y.; Qiu, Y.; Wang, J.; Liu, Y.; Wei, Y.; et al. Epidemiological and clinical characteristics of 99 cases of 2019 novel coronavirus pneumonia in Wuhan, China: A descriptive study. *Lancet* **2020**, *395*, 507–513. [CrossRef]
4. Xiang, Y.T.; Yang, Y.; Li, W.; Zhang, L.; Zhang, Q.; Cheung, T.; Ng, C.H. Timely mental health care for the 2019 novel coronavirus outbreak is urgently needed. *Lancet Psychiatry* **2020**, *7*, 228–229. [CrossRef]
5. Calabrò, R.S.; Maggio, M.G. Telepsychology: A new way to deal with relational problems associated with the COVID-19 epidemic. *Acta Biomed.* **2020**, *91*, e2020140. [PubMed]
6. De Luca, R.; Calabrò, R.S. How the COVID-19 Pandemic is Changing Mental Health Disease Management: The Growing Need of Telecounseling in Italy. *Innov. Clin. Neurosci.* **2020**, *17*, 16–17. [PubMed]
7. Calabrò, R.S.; Manuli, A.; Naro, A.; Rao, G. How Covid 19 has changed Neurorehabilitation in Italy: A critical appraisal. *Acta Biomed.* **2020**, *10*, e2020143.
8. Cuffaro, L.; Di Lorenzo, F.; Bonavita, S.; Tedeschi, G.; Leocani, L.; Lavorgna, L. Dementia care and COVID-19 pandemic: A necessary digital revolution. *Neurol. Sci.* **2020**, *41*, 1977–1979. [CrossRef]
9. Wang, J.; Xiao, L.D.; He, G.P.; De Bellis, A. Family caregiver challenges in dementia care in a country with undeveloped dementia services. *J. Adv. Nurs.* **2014**, *70*, 1369–1380. [CrossRef]
10. Christensen, M.D.; White, H.K. Dementia assessment and management. *J. Am. Med. Dir. Assoc.* **2007**, *8*, e89–e98. [CrossRef]
11. De Cola, M.C.; Lo Buono, V.; Mento, A.; Foti, M.; Marino, S.; Bramanti, P.; Manuli, A.; Calabro, R.S. Unmet Needs for Family Caregivers of Elderly People With Dementia Living in Italy: What Do We Know So Far and What Should We Do Next? *Inquiry* **2017**, *54*, 0046958017713708. [CrossRef] [PubMed]
12. Nobili, G.; Massaia, M.; Isaia, G.C.; Cappa, G.; Pilon, S.; Mondino, S.; Isaia, G.C. Valutazione dei bisogni del caregiver di pazienti affetti da demenza: Esperienza in una unità di valutazione Alzheimer. *G. Gerontol.* **2011**, *59*, 71–74.
13. World Health Organization. *The Epidemiology and Impact of Dementia: Current State and Future Trends*; Report No.: WHO/MSD/MER/15.3; WHO: Geneva, Switzerland, 2015. Available online: http://www.who.int/mental_health/neurology/dementia/dementia_thematicbrief_epidemiology.pdf (accessed on 10 September 2021).
14. Etters, L.; Goodall, D.; Harrison, B.E. Caregiver burden among dementia patient caregivers: A review of the literature. *J. Am. Acad. Nurse Pract.* **2008**, *20*, 423–428. [CrossRef] [PubMed]
15. Chiao, C.Y.; Wu, H.S.; Hsiao, C.Y. Caregiver burden for informal caregivers of patients with dementia: A systematic review. *Int. Nurs. Rev.* **2015**, *62*, 340–350. [CrossRef] [PubMed]
16. Hardan, L.; Filtchev, D.; Kassem, R.; Bourgi, R.; Lukomska-Szymanska, M.; Tarhini, H.; Salloum-Yared, F.; Mancino, D.; Kharouf, N.; Haikel, Y. COVID-19 and Alzheimer's Disease: A Literature Review. *Medicina* **2021**, *57*, 1159. [CrossRef]

17. Cipriani, G.; Di Fiorino, M.; Cammisuli, D.M. Dementia in the era of COVID-19. Some considerations and ethical issues. *Psychogeriatrics* **2021**. [CrossRef]
18. Mondo, M.; Sechi, C.; Cabras, C. Psychometric evaluation of three versions of the Italian perceived stress scale. *Curr. Psychol.* **2021**, *40*, 1884–1892. [CrossRef]
19. Sica, C.; Magni, C.; Ghisi, M.; Altoè, G.; Sighinolfi, C.; Chiri, L.R.; Franceschini, S. Coping Orientation to Problems Experienced-Nuova Versione Italiana (COPE-NVI): Uno strumento per la misura degli stili di coping. *Psicoter. Cogn. Comport.* **2008**, *14*, 27–53.
20. Novak, M.; Guest, C. Caregiver Burden Inventory (CBI). *Gerontologist* **1989**, *29*, 798–803. [CrossRef] [PubMed]
21. Ware, J.; Kosinski, M.; Keller, S.D. A 12-Item Short-Form Health Survey: Construction of scales and preliminary tests of reliability and validity. *Med. Care* **1996**, *34*, 220–233. [CrossRef]
22. Marvardi, M.; Mattioli, P.; Spazzafumo, L.; Mastriforti, R.; Rinaldi, P.; Polidori, M.C.; Cherubini, A.; Quartesan, R.; Bartorelli, L.; Bonaiuto, S. The Caregiver Burden Inventory in evaluating the burden of caregivers of elderly demented patients: Results from a multicenter study. *Aging Clin. Exp. Res.* **2005**, *17*, 46–53. [CrossRef]
23. Brown, E.E.; Kumar, S.; Rajji, T.K.; Pollock, B.G.; Mulsant, B.H. Anticipating and Mitigating the Impact of the COVID-19 Pandemic on Alzheimer's Disease and Related Dementias. *Am. J. Geriatr. Psychiatry* **2020**, *28*, 712–721. [CrossRef] [PubMed]
24. Simonetti, A.; Pais, C.; Jones, M.; Cipriani, M.C.; Janiri, D.; Monti, L.; Landi, F.; Bernabei, R.; Liperoti, R.; Sani, G. Neuropsychiatric Symptoms in Elderly With Dementia During COVID-19 Pandemic: Definition, Treatment, and Future Directions. *Front. Psychiatry* **2020**, *11*, 579842. [CrossRef]
25. Baschi, R.; Luca, A.; Nicoletti, A.; Caccamo, M.; Cicero, C.E.; D'Agate, C.; Di Giorgi, L.; La Bianca, G.; Castro, T.L.; Zappia, M.; et al. Changes in Motor, Cognitive, and Behavioral Symptoms in Parkinson's Disease and Mild Cognitive Impairment During the COVID-19 Lockdown. *Front. Psychiatry* **2020**, *11*, 590134. [CrossRef]
26. Isik, A.T.; Soysal, P.; Solmi, M.; Veronese, N. Bidirectional relationship between caregiver burden and neuropsychiatric symptoms in patients with Alzheimer's disease: A narrative review. *Int. J. Geriatr. Psychiatry* **2019**, *34*, 1326–1334. [CrossRef] [PubMed]
27. Liu, Z.; Sun, Y.Y.; Zhong, B.L. Mindfulness-based stress reduction for family carers of people with dementia. *Cochrane Database Syst Rev.* **2018**, *8*, CD012791. [CrossRef] [PubMed]
28. Pinquart, M.; Sörensen, S. Differences between caregivers and non-caregivers in psychological health and physical health: A meta-analysis. *Psychol. Aging* **2003**, *18*, 250–267. [CrossRef]
29. Roepke, S.K.; Allison, M.; von Känel, R.; Mausbach, B.; Chattillion, E.A.; Harmell, A.L.; Patterson, T.L.; Dimsdale, J.E.; Mills, P.J.; Ziegler, M.G.; et al. Relationship between chronic stress and carotid intima-media thickness (IMT) in elderly Alzheimer's disease caregivers. *Stress* **2012**, *15*, 121–129. [CrossRef]
30. Manini, A.; Brambilla, M.; Maggiore, L.; Pomati, S.; Pantoni, L. The impact of lockdown during SARS-CoV-2 outbreak on behavioral and psychological symptoms of dementia. *Neurol. Sci.* **2021**, *42*, 825–833. [CrossRef]
31. Barguilla, A.; Fernández-Lebrero, A.; Estragués-Gázquez, I.; García-Escobar, G.; Navalpotro-Gómez, I.; Manero, R.M.; Puente-Periz, V.; Roquer, J.; Puig-Pijoan, A. Effects of COVID-19 Pandemic Confinement in Patients With Cognitive Impairment. *Front. Neurol.* **2020**, *11*, 589901. [CrossRef]
32. Alexopoulos, P.; Soldatos, R.; Kontogianni, E.; Frouda, M.; Aligianni, S.L.; Skondra, M.; Passa, M.; Konstantopoulou, G.; Stamouli, E.; Katirtzoglou, E.; et al. COVID-19 Crisis Effects on Caregiver Distress in Neurocognitive Disorder. *J. Alzheimer's Dis.* **2021**, *79*, 459–466. [CrossRef]
33. Altieri, M.; Santangelo, G. The Psychological Impact of COVID-19 Pandemic and Lockdown on Caregivers of People with Dementia. *Am. J. Geriatr. Psychiatry* **2021**, *29*, 27–34. [CrossRef] [PubMed]
34. Mazzi, M.C.; Iavarone, A.; Musella, C.; De Luca, M.; de Vita, D.; Branciforte, S.; Coppola, A.; Scarpa, R.; Raimondo, S.; Sorrentino, S.; et al. Time of isolation, education and gender influence the psychological outcome during COVID-19 lockdown in caregivers of patients with dementia. *Eur. Geriatr. Med.* **2020**, *11*, 1095–1098. [CrossRef]
35. Boutoleau-Bretonnière, C.; Pouclet-Courtemanche, H.; Gillet, A.; Bernard, A.; Deruet, A.-L.; Gouraud, I.; Lamy, E.; Mazoué, A.; Rocher, L.; Bretonnière, C.; et al. Impact of Confinement on the Burden of Caregivers of Patients with the Behavioral Variant of Frontotemporal Dementia and Alzheimer Disease during the COVID-19 Crisis in France. *Dement. Geriatr. Cogn. Disord. Extra* **2020**, *10*, 127–134. [CrossRef] [PubMed]
36. Rotondo, E.; Galimberti, D.; Mercurio, M.; Giardinieri, G.; Forti, S.; Vimercati, R.; Borracci, V.; Fumagalli, G.G.; Pietroboni, A.M.; Carandini, T.; et al. Caregiver Tele-Assistance for Reduction of Emotional Distress During the COVID-19 Pandemic. Psychological Support to Caregivers of People with Dementia: The Italian Experience. *J. Alzheimer's Dis.* **2021**, *69*, 1–8. [CrossRef]
37. Godwin, K.M.; Mills, W.L.; Anderson, J.A.; Kunik, M.E. Technology-driven interventions for caregivers of persons with dementia: A systematic review. *Am. J. Alzheimer's Dis. Other Demen.* **2013**, *28*, 216–222. [CrossRef] [PubMed]
38. Heckman, J.J. Selection Bias and Self-selection. In *Econometrics*; Eatwell, J., Milgate, M., Newman, P., Eds.; Palgrave Macmillan: London, UK, 1990.

Journal of
Clinical Medicine

Review

Autopsy Findings and Causality Relationship between Death and COVID-19 Vaccination: A Systematic Review

Francesco Sessa [1], Monica Salerno [2], Massimiliano Esposito [2], Nunzio Di Nunno [3], Paolo Zamboni [4] and Cristoforo Pomara [2,*]

1 Department of Clinical and Experimental Medicine, University of Foggia, 71122 Foggia, Italy; francesco.sessa@unifg.it
2 Department of Medical, Surgical and Advanced Technologies "G.F. Ingrassia", University of Catania, 95121 Catania, Italy; monica.salerno@unict.it (M.S.); massimiliano.esposito91@gmail.com (M.E.)
3 Department of History, Society and Studies on Humanity, University of Salento, 73100 Lecce, Italy; nunzio.dinunno@unisalento.it
4 Vascular Diseases Center, Hub Center for Venous and Lymphatic Diseases Regione Emilia-Romagna, Sant'Anna University Hospital of Ferrara, 44121 Ferrara, Italy; zambo@unife.it
* Correspondence: cristoforo.pomara@unict.it; Tel.: +39-095-3782-153 or +39-333-2466-148

Abstract: The current challenge worldwide is the administration of the severe acute respiratory syndrome coronavirus 2 (SARS-CoV-2) vaccine. Considering that the COVID-19 vaccination represents the best possibility to resolve this pandemic, this systematic review aims to clarify the major aspects of fatal adverse effects related to COVID-19 vaccines, with the goal of advancing our knowledge, supporting decisions, or suggesting changes in policies at local, regional, and global levels. Moreover, this review aims to provide key recommendations to improve awareness of vaccine safety. All studies published up to 2 December 2021 were searched using the following keywords: "COVID-19 Vaccine", "SARS-CoV-2 Vaccine", "COVID-19 Vaccination", "SARS-CoV-2 Vaccination", and "Autopsy" or "Post-mortem". We included 17 papers published with fatal cases with post-mortem investigations. A total of 38 cases were analyzed: 22 cases were related to ChAdOx1 nCoV-19 administration, 10 cases to BNT162b2, 4 cases to mRNA-1273, and 2 cases to Ad26.COV2.S. Based on these data, autopsy is very useful to define the main characteristics of the so-called vaccine-induced immune thrombotic thrombocytopenia (VITT) after ChAdOx1 nCoV-19 vaccination: recurrent findings were intracranial hemorrhage and diffused microthrombi located in multiple areas. Moreover, it is fundamental to provide evidence about myocarditis related to the BNT162B2 vaccine. Finally, based on the discussed data, we suggest several key recommendations to improve awareness of vaccine safety.

Keywords: COVID-19 vaccination; fatal case; adverse events following immunization (AEFI); vaccine-induced immune thrombotic thrombocytopenia (VITT)

Citation: Sessa, F.; Salerno, M.; Esposito, M.; Di Nunno, N.; Zamboni, P.; Pomara, C. Autopsy Findings and Causality Relationship between Death and COVID-19 Vaccination: A Systematic Review. *J. Clin. Med.* **2021**, *10*, 5876. https://doi.org/10.3390/jcm10245876

Academic Editor: Giorgio I. Russo

Received: 2 December 2021
Accepted: 10 December 2021
Published: 15 December 2021

Publisher's Note: MDPI stays neutral with regard to jurisdictional claims in published maps and institutional affiliations.

1. Introduction

Severe acute respiratory syndrome coronavirus 2 (SARS-CoV-2) has been identified as the causative agent of coronavirus disease 2019 (COVID-19) [1]. The virus rapidly spread around the world, leading to one of the most severe pandemics in human history: on 2 December 2021, there were 263,565,559 cases worldwide, with more than 5,225,667 confirmed deaths, affecting 223 countries [2]. Vaccination is undoubtedly the most effective tool for preventing infectious diseases, representing one of the most important breakthroughs in the history of medical science. To date, more than 4.29 billion vaccine doses have been administered worldwide, reaching about 55.9% of the global population. About 74% of these vaccinations have been administered in high- and upper-middle-income countries, and only 0.8% in low-income countries. In this context, it is important to note that several high-income countries are starting to receive an additional dose, while in low-income countries the number of fully vaccinated people is alarmingly low [3].

The development of COVID-19 vaccines started in January 2020 with the identification of the genetic sequence of SARS-CoV-2. Subsequently, many vaccine candidates were tested for the development of safe and effective COVID-19 vaccines, exploring different technologies such as mRNA, subunit proteins, and virus-based vaccines such as inactivated, live-attenuated, and recombinant viral vaccines. A new COVID-19 vaccine that uses circular strands of DNA to prime the immune system has recently been approved [4]. COVID-19 vaccination has been found to reduce the risk of SARS-CoV-2 transmission as well as hospitalization and associated complications [5]. This is attributed to vaccine efficacy through its ability to induce both humoral and cell-mediated immune responses in vaccinated subjects [6]. It has recently been reported that vaccines averted over one thousand deaths in Israeli during the first 4-months of the vaccination campaign [7]. Moreover, almost all patients hospitalized with COVID-19 by the end of May in Polish hospitals were not vaccinated [8]. Similar data have been published worldwide.

Following the recommendation of the European Medicines Agency (EMA), the European Union authorized the use of four vaccines, opening the way to a gradual return to pre-pandemic life. In particular, the vaccine BNT162b2 (Pfizer–BioNTech) was authorized on 21 December 2020 [9]; another vaccine, mRNA-1273 (Moderna), was approved on 6 January 2021 [10]; the third vaccine, ChAdOx1 nCov-19 (AstraZeneca), was approved on 29 January 2021 [11]; the fourth vaccine is Ad26.COV2.S (COVID-19 Vaccine Janssen—Johnson & Johnson), authorized on 11 March 2021 [12]. At the moment of writing, only these vaccines have been authorized for use in the European Union, while Sputnik V (Gam-COVID-Vac), COVID-19 Vaccine (Vero Cell), inactivated, and Vidprevtyn are currently under rolling review; finally, Nuvaxovid (also known as NVX-CoV2373) started the process of marketing authorization on 16 November 2021 [13].

It is important to note that the evaluation of adverse events following vaccination is a pivotal part of the clinical trials conducted pre-authorization. Moreover, clinical trials are not designed to detect very rare adverse events; this requires post-authorization monitoring. For example, in Europe, as reported on the EMA website, clinical trials are usually conducted on carefully selected patients and followed up very closely under controlled conditions. This means that at the time a medicine is authorized, as well as a vaccine, it has been tested on a relatively small number of selected patients for a limited length of time. For these reasons, it is essential that all medicines are monitored for safety throughout their use in healthcare practice. In this way, a pharmacovigilance system is mandatory after drug approval, monitoring suspected adverse reactions [14].

However, from a public health viewpoint, several important issues are still present and relevant in COVID-19 vaccines. They regard not only vaccine efficacy and protection duration, but also safety. In various countries, severe and fatal adverse effects occurring at the same time as COVID-19 vaccination have been reported [15], generating hesitancy and suspicion in the population [16,17].

Considering that COVID-19 vaccination represents the best possibility to resolve this the pandemic, this systematic review aims to clarify the major aspects of fatal adverse effects related to COVID-19 vaccines, with a further goal of advancing our knowledge supporting decisions or suggesting changes in policies at local, regional, and global levels. Moreover, this review aims to provide key recommendations to improve awareness of vaccine safety.

2. Materials and Methods

2.1. Search Strategy

The Preferred Reporting Items for Systematic Reviews and Meta-Analyses Statement (PRISMA) recommendations were applied to perform this systematic review [18]. We searched all publications related to SARS-CoV-2 vaccines and fatal adverse effects from the following databases: Scopus, EMBASE, Medline (via PubMed), and Web of Science. All studies published up to 2 December 2021 were searched without language restriction by three independent reviewers. Searched medical subject headings (MeSH) were: "COVID-19

Vaccine", "SARS-CoV-2 Vaccine", "COVID-19 Vaccination", "SARS-CoV-2 Vaccination", and "Autopsy" or "Post-mortem". References and citation lists of selected articles and reviews were also reviewed for any other relevant literature.

2.2. Study Selection

The retrieved studies were first reviewed by three independent authors based on the title and abstract (FS, MS, and ME), all unrelated publications were removed, and the full texts of the remaining articles were fully reviewed. Then, two independent reviewers (CP and PZ) judged potentially eligible articles, and disagreements were resolved by discussion and for each article a consensus was reached.

2.3. Eligibility, Inclusion, and Exclusion Criteria

The following predetermined conditions had to be met for studies to be considered for inclusion in this meta-analysis. For initial screening, all studies with post-mortem investigations were included in the systematic review; English language was an inclusion criterion. All papers without post-mortem investigations, review articles, and studies with no extractable data were excluded from this review.

3. Results

A total of 53 publications matched the research parameters; removing duplicates, 33 articles were fully screened for COVID-19 vaccines and fatal adverse events. Out of these studies, 17 met the systematic review inclusion criteria (Figure 1).

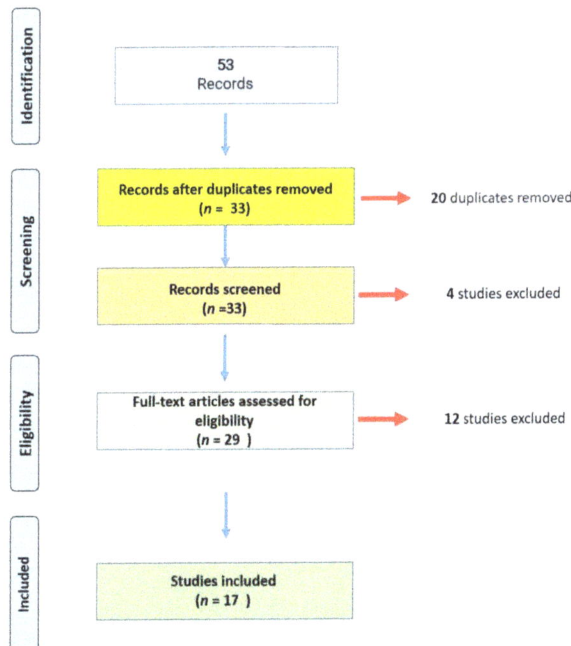

Figure 1. Flow diagram of the literature search and study selection for this systematic review (PRISMA flow chart).

As summarized in Table 1, 17 papers were published with fatal cases occurring at the same time as COVID-19 vaccine administration. A total of 38 cases (19 females, 19 males) were described: 22 cases were related to ChAdOx1 nCoV-19 administration, 10 cases to BNT162b2, 4 cases to mRNA-1273, and 2 cases to Ad26.COV2.S (Janssen).

Table 1. The main data of the post-mortem examination performed on cases occurring at the same time as COVID-19 vaccination: on a total of 38 cases, 22 patients were vaccinated with ChAdOx1 nCoV-19, 10 cases with BNT162b2, 4 cases with mRNA-1273, 2 cases with Ad26.COV2.S.

Reference	Vaccine	Fatal Cases				Post-Mortem Findings	Causality Relationship	WHO Algorithm
		Sex, Age	D	H	R			
Greinacher et al. [19]	ChAdOx1 nCoV-19	M, 49 y.o.	10	1	n.a.	Cerebral venous thrombosis; portal-vein thrombosis, including the splenic and upper mesenteric veins; in addition, small thrombi were visualized in the infrarenal aorta and both iliac arteries.	YES	NOT DESCRIBED
Althaus et al. [20]	ChAdOx1 nCoV-19	F, 48 y.o.	6	10	n.a.	Complete thrombotic obstruction of the straight, sagittal and transversal cerebral sinuses; subarachnoid hemorrhage; cerebral edema and bilateral pulmonary embolism; obstruction of glomerular arterioles and capillaries by hyaline microthrombi containing fibrin and platelets.	YES	NOT DESCRIBED
		M, 24 y.o.	10	7	Het. FVL	Massive cerebral hemorrhage and cerebral edema, bilateral pulmonary thromboembolism and obstruction of glomeruli by hyaline microthrombi.	YES	NOT DESCRIBED
Mauriello et al. [21]	ChAdOx1 nCoV-19	F, 48 y.o.	18	21	pre-existing condition of thrombocytopenia due to myelodysplasia	Massive cerebral hemorrhage; purulent abscess involving the right fronto-temporo-parietal lobes, the nucleus of the right base, with midline shift and wedging of the cerebellar tonsils and an internal and external hemotocephalus.	YES	NOT DESCRIBED
Wieldmann et al. [22]	ChAdOx1 nCoV-19	F, 34 y.o.	7	1	None	Edematous brain with sparse subarachnoid hemorrhage and a large hemorrhagic infarction in the right hemisphere; thrombi were present in both transverse sinuses.	YES	NOT DESCRIBED
	ChAdOx1 nCoV-19	F, 42 y.o.	10	15	n.a.	Thrombus in the left transverse and sigmoid sinus, as well as in the sagittal cerebral sinus; massive hemorrhagic infarction in the left hemisphere; peripheral areas with infarction in the lungs.	YES	NOT APPLIED
	ChAdOx1 nCoV-19	F, 37 y.o.	8	3	n.a.	Large hemorrhagic infarction in the left cerebral hemisphere; extensive hemorrhagic changes in the cerebellum, as well as focal white substance hemorrhages in the cerebral hemispheres and in the brainstem. Thrombi were present in the left transverse and sigmoid sinuses.	YES	NOT DESCRIBED
	ChAdOx1 nCoV-19	F, 54 y.o.	6	2	n.a.	Thrombi in the posterior sagittal sinus and both transverse sinuses. Massive hemorrhagic venous infarction in the right parietal lobe and bilateral hemorrhagic infarctions in multiple cortical areas.	YES	NOT DESCRIBED

Table 1. *Cont.*

Reference	Vaccine	Fatal Cases				Post-Mortem Findings	Causality Relationship	WHO Algorithm
		Sex, Age	D	H	R			
Bjørnstad-Tuveng et al. [23]	ChAdOx1 nCoV-19	F, n.a. (young)	7	n.a.	None	Intracranial hemorrhage. Moreover, small thrombi were found in the transverse sinus, frontal lobe, and pulmonary artery.	YES	NOT DESCRIBED
Scully et al. [24]	ChAdOx1 nCoV-19	F, 55 y.o.	6	n.a.	n.a.	Thrombosis in many small vessels, especially vessels in the lungs and intestine, cerebral veins, and venous sinuses, as well as evidence of extensive intracerebral hemorrhage.	YES	NOT DESCRIBED
Günther et al. [25]	ChAdOx1 nCoV-19	M, 54 y.o.	12	1	None	Residual thrombus in the left sinus transversus; no evidence for other thromboembolic pathology in the brain or other solid organs was found.	YES	NOT DESCRIBED
Pomara et al. [26,27]	ChAdOx1 nCoV-19	M, 50 y.o.	10	6	None	Portal and mesenteric thrombosis with extension into the splenic vein. Moreover, extensive cerebral hemorrhages were described.	YES	YES
		F, 37 y.o.	13	10	None	Thrombi in cerebral sinus; massive thrombosis of the whole venous tree of left upper limb extending from the hand to the axillary vein, with symmetric lesions in the veins of the right hand and the right axillary vein.	YES	YES
Schneider et al. [28]	ChAdOx1 nCoV-19	F, 32 y.o.	12	Home	None	Massive cerebral hemorrhage, anti-PF4 heparin antibody tests: positive, HIPA-Test: positive, PIPA-Test: positive.	Very likely	NOT DESCRIBED
	ChAdOx1 nCoV-19	F, 34 y.o.	1	Home	Obesity, massive cardiac hypertrophy, myocardial infarction scars	Recurrent myocardial infarction in the presence of massive cardiac hypertrophy.	NO	NOT DESCRIBED
	ChAdOx1 nCoV-19	F, 48 y.o.	10	Workplace	None	Aortic dissection with rupture, high blood loss.	NO	NOT DESCRIBED
	ChAdOx1 nCoV-19	M, 63 y.o.	14	Home	Severe pre-existing cardiac changes	Severe coronary sclerosis, cardiac hypertrophy, myocardial infarction scars, liver cirrhosis.	NO	NOT DESCRIBED
	ChAdOx1 nCoV-19	M, 61 y.o.	1	Home	Severe pre-existing cardiac changes	Severe coronary sclerosis, massive cardiac hypertrophy, negative anaphylaxis diagnostics.	NO	NOT DESCRIBED
	ChAdOx1 nCoV-19	M, 71 y.o.	10	Home	Severe coronary sclerosis, massive cardiac hypertrophy, myocardial infarction scars	Pulmonary embolism in the presence of Deep Vein Thrombosis.	NO	NOT DESCRIBED

Table 1. *Cont.*

Reference	Vaccine	Fatal Cases				Post-Mortem Findings	Causality Relationship	WHO Algorithm
		Sex, Age	D	H	R			
	ChAdOx1 nCoV-19 (second dose)	F, 38 y.o.	8	Hospital (n.a.)	n.a.	Multiple fresh thrombi, including in the cerebral venous sinuses, cardiac hypertrophy, fresh myocardial infarction, hypoxic brain damage, anti-PF4 heparin antibody tests: positive, HIPA-Test: positive, PIPA-Test: positive.	Unlikely	NOT DESCRIBED
	ChAdOx1 nCoV-19	F, 65 y.o.	10	Hospital (n.a.)	n.a.	Signs of a bleeding diathesis, cerebral hemorrhages, CVT, mild coronary sclerosis, anti-PF4 heparin antibody tests: positive, HIPA-Test: positive, PIPA-Test: positive.	Very likely	NOT DESCRIBED
	ChAdOx1 nCoV-19	M, 57 y.o.	2	Hospital (n.a.)	Massive cardiac hypertrophy	Severe coronary sclerosis, extensive myocardial infarction scars, fresh myocardial infarction.	NO	NOT DESCRIBED
		F, n.a. (elderly)	5	0	Coronary heart disease, cardiac insufficiency, arterial hypertension, dementia and hyperthyroidism.	Pulmonary artery embolism with infarction of the right lower lobe of the lung with deep leg vein thromboses on both sides.	NO	NOT DESCRIBED
			She was found dead					
Edler et al. [29]	BNT162b2	M, n.a. (elderly)	10	2	Chronic renal failure, anemia, atrial fibrillation, pulmonary artery embolism, arterial hypertension, peripheral artery disease, right thalamic infarction with left hemiparesis, recurrent tonic-clonic seizures, gait disorder with polyneuropathy, rheumatoid arthritis and prostate carcinoma with prostatectomy.	Nasopharyngeal swab for SARS-CoV-2 RNA was positive. Autopsy revealed chronic and acute pancreatitis. Pneumonia was confirmed as the cause of death.	NO	NOT DESCRIBED
		M, n.a.	2 (he was found dead)	0	Apoplexy and myocardial infarction as well as arterial hypertension and type II diabetes mellitus.	The known pre-existing conditions were confirmed, and further organ pathologies typical of old age were found in the form of signs of chronic obstructive pulmonary disease (COPD) and chronic renal dysfunction.	NO	NOT DESCRIBED
Hansen et al. [30]	BNT162b2	M, 86 y.o.	18	7	Past medical history included systemic arterial hypertension, chronic venous insufficiency, dementia and prostate carcinoma.	Nasopharyngeal swab for SARS-CoV-2 RNA was positive (day 24). No characteristic morphological features of COVID-19 were reported (i.e., alveolar damage in the lungs); extensive acute bronchopneumonia, possibly of bacterial origin.	NO	NOT DESCRIBED

Table 1. *Cont.*

Reference	Vaccine	Fatal Cases				Post-Mortem Findings	Causality Relationship	WHO Algorithm
		Sex, Age	D	H	R			
Schneider et al. [28]	BNT162b2	M, 65 y.o.	1	Home	Severe pre-existing cardiac changes	Severe coronary sclerosis, massive cardiac hypertrophy, myocardial infarction scars, myocarditis.	POSSIBLE	NOT DESCRIBED
	BNT162b2	M, 71 y.o.	1	Home	Severe pre-existing cardiac changes	Massive cardiac hypertrophy, coronary sclerosis, negative anaphylaxis diagnostics.	NO	NOT DESCRIBED
	BNT162b2	F, 72 y.o.	12	Home	Coronary sclerosis, cardiac hypertrophy	Massive cerebral hemorrhage.	NO	NOT DESCRIBED
	BNT162b2 (second dose)	M, 79 y.o.	6	Home	Deep Vein Thrombosis	Massive pulmonary embolism, coronary sclerosis, pericarditis, chronic pulmonary Emphysema.	NO	NOT DESCRIBED
	BNT162b2 (second dose)	F, 72 y.o.	0	Vaccination center	n.a.	Severe coronary sclerosis with coronary thrombosis, myocardial infarction scars, fresh myocardial infarction.	NO	NOT DESCRIBED
Choi et al. [31]	BNT162b2	M, 22 y.o.	5	7 h	None	On microscopic examination, diffuse inflammatory infiltration, with neutrophil and histiocyte predominance was observed within the myocardium. Notably, the inflammatory infiltrates were dominant in the atria, and around the sinoatrial (SA) and atrioventricular (AV) nodes, whereas the ventricular area displayed minimal or no inflammatory cells. Occasional myocyte necrosis or degeneration was found adjacent to the inflammatory infiltrates, without abscess formation or bacterial colonization. The cause of death was determined to be myocarditis.	YES	NOT DESCRIBED
Verma et al. [32]	mRNA-1273 (second dose)	M, 42 y.o.	15	7	None	An inflammatory infiltrate admixed with macrophages, T-cells, eosinophils, and B cells was observed in heart tissue. The cause of death was defined as fulminant myocarditis that had developed within 2 weeks after COVID-19 vaccination.	YES	NOT DESCRIBED

Table 1. *Cont.*

Reference	Vaccine	Fatal Cases				Post-Mortem Findings	Causality Relationship	WHO Algorithm
		Sex, Age	D	H	R			
Schneider et al. [28]	mRNA-1273	M, 82 y.o.	1	Home	Pre-existing cardiac changes with infarction	Severe coronary sclerosis, massive cardiac hypertrophy, extensive myocardial infarction scars, negative anaphylaxis diagnostics.	NO	NOT DESCRIBED
	mRNA-1273	F, 91 y.o.	1	Home	Pre-existing cardiac changes with infarction	Severe coronary sclerosis, massive cardiac hypertrophy, myocardial infarction scars, negative anaphylaxis diagnostics.	NO	NOT DESCRIBED
	mRNA-1273 (second dose)	F, 57 y.o.	6.	Home.	Hyperglycemic coma	Severe coronary sclerosis, fatty liver, high levels of glucose and lactate in cerebrospinal fluid (CSF) and aqueous humor exceeding the cumulative levels of Traub.	NO	NOT DESCRIBED
Schneider et al. [28]	Ad26.COV2.S (Janssen)	M, 69 y.o.	9	Home	n.a.	CVT, severe coronary sclerosis with coronary thrombosis, massive cardiac hypertrophy, fresh myocardial infarction, anti-PF4 heparin antibody tests: positive, HIPA-Test: positive, PIPA-Test: positive.	POSSIBLE	NOT DESCRIBED
Choi et al. [33]	Ad26.COV2.S (Janssen)	M, 38 y.o.	2	10 h	Smoldering multiple myeloma had been diagnosed 1.5 years before	Autopsy results showed no evidence of acute infection or cardiovascular disease in the internal organs. Moreover, pulmonary edema, pleural effusion, and pericardial effusion were reported.	POSSIBLE	NOT DESCRIBED

Legend: (D) first symptoms after vaccination (days); (H) Hospitalization (days); (R) clinical features; (n.a.) not available/performed; (FVL) Factor V Leiden; (Het.) Heterozygous.

Through a box plot analysis, we have summarized the data about the age of subjects involved in fatal adverse events after vaccination (Figure 2A), and the data about the time interval between vaccine administration and the first symptoms (Figure 2B).

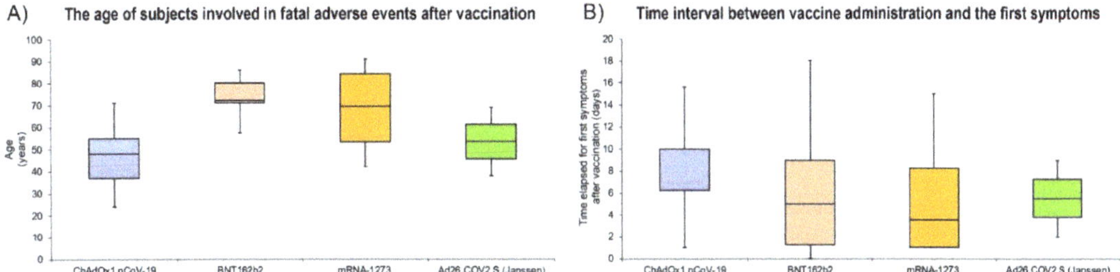

Figure 2. Box plot analysis comparing the age of subjects involved in fatal cases after vaccine administration (**A**). Comparison of the time interval between vaccine administration and the first symptoms (**B**).

Twenty-two (14 females, 8 males) cases were reported as deaths occurring at the same time as ChAdOx1 nCoV-19 administration. The mean age was 47.6 ± 12.28 (Male $= 53.6 \pm 13.9$; Female $= 44 \pm 9.9$).

Greinacher et al. [19] reported six cases of fatal adverse effects after COVID-19 administration, even if only one case is discussed in their report. Limiting the comments to this case, the authors reported a portal-vein thrombosis; moreover, they described thrombi in the splenic and upper mesenteric veins; finally, small thrombi were reported in the infrarenal aorta and both iliac arteries. Finally, autopsy findings revealed cerebral venous thrombosis. This paper described, for the first time, the presence of antibodies against platelet factor 4 (PF4), suggesting a similar pathological mechanism to severe heparin-induced thrombocytopenia (HIT).

Althaus et al. [20] discussed eight cases of death after ChAdOx1 nCoV-19 administration, although the post-mortem examination was performed in only two cases, as reported in Table 1. The main findings were massive cerebral hemorrhage with edema, and bilateral pulmonary thromboembolism. In both cases, the authors reported the presence of microthrombi on glomeruli. The authors concluded that all patients developed vaccine-induced immune thrombotic thrombocytopenia (VITT) after the administration of SARS-CoV-2 vaccine ChAdOx1 nCoV-19. This diagnosis was based on the presence of a high antibody titer against PF4. In this paper, the authors suggested that the presence of PF4 antibodies in VITT patients induced a significant increase in procoagulant markers.

Mauriello et al. [21] presented a fatal case of thromboembolism following administration of the first dose of ChAdOx1 nCOV-19 (AstraZeneca). At autopsy, massive cerebral hemorrhage was found, even if the level of serum anti-PF4 antibodies was undetectable. Based on their report, the authors suggested avoiding the use of ChAdOx1 nCOV-19 vaccine in subjects with a pre-existing condition of thrombocytopenia due to myelodysplasia, such as in the reported case.

Wieldmann et al. [22] presented a case series of five women with rapid progressive neurological symptoms, cerebral venous thrombosis (CVT) with intracerebral hemorrhage and thrombocytopenia, occurring 7/10 days after ChAdOx1 nCOV-19 (AstraZeneca) vaccination. Four of them died and autopsies were performed. The post-mortem findings are very similar in all subjects involved: cerebral hemorrhage with the presence of thrombi at the level of the sinuses. In all cases, the authors reported the presence of anti-PF4 antibodies.

Bjørnstad-Tuveng et al. [23] discussed a single case of a female healthcare worker who died of intracranial hemorrhage. Moreover, the authors described the presence of small thrombi in the transverse sinus, frontal lobe, and pulmonary artery. In light of the previous studies, the authors performed the anti-PF4 tests confirming the presence of these antibodies.

Scully et al. [24] reported seven cases, even if the post-mortem examination was performed in only one case, describing evidence of thrombosis in many small vessels located in the lungs, intestine, cerebral veins, and venous sinuses. Moreover, an extensive intracerebral hemorrhage and positivity for the anti-PF4 test were reported.

Günther et al. [25] described the case of a subject who presented with typical symptoms of VITT, including thrombocytopenia, cerebral venous and sinus thrombosis (CVST), and signs of disseminated intravascular coagulation (DIC). The presence of anti-PF4 antibodies was reported. The post-mortem findings confirmed the presence of residual thrombus in the left sinus transversus without evidence in the brain or in other organs.

Pomara et al. [26] presented two cases (one male and one female) of death after vaccine administration: the presence of extensive cerebral hemorrhages was reported in both cases. Moreover, in one case, portal and mesenteric thromboses with extension into the splenic vein were described, while, in the other case, massive thrombosis of the whole venous tree of the left upper limb extending from the hand to the axillary vein, with symmetric lesions in the veins of the right hand and the right axillary vein, was reported. In both cases, the anti-PF4 test was positive. It is important to note that for the first time the causality WHO algorithm was adopted to determine the direct link between vaccination and a fatal adverse effect [27]. Moreover, the same group suggested inserting autopsy as an essential tool that should be carried out in each suspected case.

Schneider et al. [28] discussed nine cases occurring at the same time as ChAdOx1 nCOV-19 vaccination: although they did not describe the application of the WHO algorithm to ascertain the causality relationship, the authors excluded it in one case, while they classified another case as "unlikely", and the other two cases as "very likely".

The fatal cases related to the BNT162b2 vaccine administration involved 10 subjects (7 females, 3 males), with an average age of 66.7 ± 20.8.

Edler et al. [29] described three cases of elderly subjects affected by severe cardiovascular diseases and other comorbidities (see Table 1). All subjects died in the context of these pre-existing conditions, while one case, testing positive at nasopharyngeal swab, developed COVID-19 pneumonia. In this report, it is important to note the pivotal role of autopsy in order to exclude a causality relationship between vaccine administration and death.

Hansel et al. [30] reported a case of an elderly male subject who had received the first dose of the BNT162b2 mRNA COVID-19 vaccine. The man was affected by several comorbidities, and although he did not present with any COVID-19-specific symptoms, he tested positive for SARS-CoV-2 before he died. The authors did not confirm the causality relationship.

Schneider et al. [28] discussed the data of five cases occurring at the same time as BNT162b2 mRNA COVID-19 vaccine administration. Based on pre-existing diseases and post-mortem findings they did not indicate a causal relationship with the vaccination. Only one case was classified as having a "possible" relationship with the vaccine administration.

Choi et al. [31] described a particular myocarditis related to the BNT162b2 mRNA COVID-19 vaccine, identifying histological differences from viral or immune-mediated myocarditis: indeed, the authors reported that the inflammatory infiltrates were predominantly neutrophils and histiocytes, rather than lymphocytes.

The fatal cases related to mRNA-1273 vaccine administration involved four subjects (two females, two males), with an average age of 68 ± 22.5.

Verma et al. [32] reported the first fatal case after mRNA-1273 vaccination: this is the first case related to the second rather than the first dose.

Schneider et al. [28] described three cases: the authors concluded that there was no relationship between death and vaccine administration based on the autopsy findings combined with pre-existing diseases.

The same authors reported one case related to the Ad26.COV2.S (Janssen) vaccine, reporting a possible causality relationship based on post-mortem findings. Similarly, Choi et al. [33] reported the fatal case of a subject who died two days after Ad26.COV2.S (Janssen) vaccination. Although the patient suffered from multiple myeloma diagnosed 1.5 years before, the cause of death was identified as fatal systemic capillary leak syndrome possible related to COVID-19 vaccination.

4. Discussion

Vaccination plays a key role in the pandemic war, representing a crucial measure of infection control [34,35]. At the time of writing, COVID-19 cases are sweeping Europe once again, particularly in those countries with a low rate of vaccination.

The first requirement is to ensure thorough, up-to-date, correct, and complete information on vaccines. In particular, their side effects must be publicized, including all useful information needed to interpret this properly in context [35]. Of course, in the case of the COVID-19 vaccination, the necessity of a promptly available vaccine has led to some adverse effects not being completely known. Although the rate of severe adverse effects is very low, it is important to highlight that in the first phase of vaccination, the package leaflet of each vaccine and the relative informed consent did not contain the unknown adverse effects that were added only after the first cases of severe adverse effects. It is important to remark that a pharmacovigilance system is mandatory after each drug approval, monitoring all suspected adverse reactions [14].

Based on the discussed data, a causality relationship between vaccine administration and death was demonstrated in 13 cases of ChAdOx1 nCOV-19 (AstraZeneca) vaccination, while it was excluded in the other 6 cases; in two cases the relationship was classified as "very likely", and in the last one as "unlikely". As concerns BNT162B2, of the ten cases reported in the literature, the causality relationship was established in one case, while in another case it was defined as "possible". Finally, the causality relationship was established in one case of mRNA-1273 vaccination and classified as "possible" in the two cases related to the Ad26.COV2.S (Janssen) vaccine. As recently noted in a review published by Sharifian-Dorche et al. [36], other severe adverse effects have been described related to other authorized vaccines.

Analyzing the international data, it has been reported that both vaccines based on the adenoviral-based vector (ChAdOx1 nCov-19 and Ad26.COV2.S Janssen) can cause similar adverse reactions, generating severe adverse effects such as thrombocytopenia and thrombosis in atypical locations (cerebral and/or splanchnic veins) in healthy subjects a few days following vaccination. Based on the data obtained through this literature review, these symptoms appeared 8.6 ± 4.1 days after vaccine administration. All included cases were related to the first dose administration. Nevertheless, these severe adverse effects are extremely rare: 3 to 10 cases per million. Similar complications are lower for the two messenger RNA (mRNA)-based vaccines (BNT162b2 and mRNA-1273): severe adverse effects have been estimated to occur in 0.8 to 1 case per million [37].

The disclosure of any risks involved in vaccination is an integral part of the information provided: consent may only be effectively „informed" when the risks and benefits are completely understood. In the case of ChAdOx1 nCoV-19 we certainly cannot affirm that at the time of the first administrations the possible effects, such as those found (cerebral hemorrhages and diffuse thrombosis), were fully known. In this scenario, the first administration was completed in the absence of complete information for the patient: it is possible to make a risk assessment only when all adverse effects are known, and the risks are quantified based on research findings [38,39].

It is interesting to note that the criteria for the diagnosis of vaccine-induced death have been adopted only by Pomara et al. [26,27]: the authors adopted the proposed WHO algorithm to establish direct causality, confirming a direct link between vaccine administration and fatal adverse effects. As recently remarked by Mungmunpuntipamtip and Wiwanitkit [40], the criteria to establish a direct link between vaccination and fatal adverse effects should be standardized by the international community; in this way, the post-mortem investigation represents an essential tool to confirm all the data obtained during hospitalization.

The post-mortem investigation remains the gold standard to define the exact cause of death and the related pathophysiological processes [41,42]. The COVID-19 vaccine campaign began in about December 2020, and, at the same time, monitoring of death associated with adverse effects started in all countries. Although different fatal events have been reported occurring at the same time as COVID-19 vaccine administration, only a few papers have been published describing the post-mortem findings (38 cases: 22 patients were vaccinated with ChAdOx1 nCoV-19, 10 cases with BNT162b2, 4 cases with mRNA-1273, and 2 cases with Ad26.COV2.S Janssen), as summarized in Table 1. Based on these data, autopsy is very useful to define the main characteristics of the so-called VITT after ChAdOx1 nCoV-19 vaccination: the recurrent findings were intracranial hemorrhage and diffused microthrombi located in multiple areas. In two cases [19,26] brain hemorrhage was preceded by portal and mesenteric thrombosis with extension into the splenic vein. Microscopic evaluation was reported only in one study [27], showing several vascular thrombi and hemorrhagic areas at the level of the brain. In addition, diffuse thrombi were observed in small and medium-sized vessels due to endothelial activation after an inflammatory reaction with a procoagulant process and subsequent thrombotic reaction. The same group conducted immunohistochemical investigations, revealing the expression of adhesion molecules and activated inflammatory cells in the vascular and perivascular tissues of different organs (such as heart, lung, liver, kidney, ileum, and deep veins). The inflammatory cells were found to be arranged in clusters with aggregated platelets at the endoluminal level, confirming a pro-thrombotic state.

In addition, as described by Rzymski et al. [43], different mechanisms could be related to the severe/fatal adverse effects after ChAdOx1 nCoV-19 vaccination: the possible role of antibodies against platelet factor 4 (PF4); the direct interaction between adenoviral vector and platelets; the possible cross-reactivity of antibodies against SARS-CoV-2 spike protein with PF4; the possible cross-reactivity of anti-adenovirus antibodies and PF4; the possible interaction between spike protein and platelets; the platelet expression of spike protein and subsequent immune response; the platelet expression of other adenoviral proteins and subsequent immuno-reactions. Finally, considering the small number of subjects involved in similar adverse events, it is also plausible that thrombotic thrombocytopenia after COVID-19 vaccine administration may be multifactorial, with a pivotal role played by the genotype and/or influenced by post-transcriptional events. An important piece of data that emerges from this review is related to the age of the subjects involved in the fatal cases occurring at the same time as COVID-19 vaccination: while for the other vaccines the subjects involved were over 65 y.o., in the case of ChAdOx1 nCoV-19 the average age was 47.6, suggesting that the severe adverse effects occur more frequently in subjects under 65. This finding was made analyzing the average age in female subjects involved in fatal adverse effects related to the ChAdOx1 nCoV-19 vaccination: we found an average age of 44 years, demonstrating a high rate of involvement of females under 50 y.o.

In 8/10 cases of BNT162b2 administration, through the data collected during autopsy, the authors [29,30] excluded the causality relationship considering the previous comorbidities of all involved subjects. Two cases are of interest for the scientific community: two deceased subjects tested positive for the COVID-19 infection. Although in both cases no signs of COVID-19 complications were found, these cases could be related to the SARS-CoV-2 variants that may allow the virus to escape host immunity, in particular, the immunity conferred by vaccination [44,45]. A direct relationship between vaccination and fatal adverse effects was reported by Choi et al. [31] who identified myocarditis as a cause of death: these findings were confirmed in a recent report describing an increased risk of myocarditis in subjects vaccinated with BNT162b2 [46].

Considering the data about the other mRNA vaccine (mRNA-1273), Verma et al. [32] reported the first fatal case after the second rather than first dose, although the same authors reported that a direct causal relationship cannot be definitively established because they did not perform testing for viral genomes or auto-antibodies in the tissue specimens.

Finally, the two cases related to the Ad26.COV2.S (Janssen) vaccine were classified as a "possible" relationship with the COVID-19 vaccination. Considering that the Ad26.COV2.S (Janssen) vaccine is based on a specific type of adenovirus, it is important to note that the anti-PF4 heparin antibody test was positive, similar to the ChAdOx1 nCoV-19 cases.

Although post-mortem investigations were reported in only a few cases, it is reasonable to assume that the potential causality between death and COVID-19 vaccination had been studied in a large number of post-mortem investigations for different reasons that had not been published: the fact that only 17 papers with post-mortem investigations were published does not mean that post-mortem investigations in deaths after vaccination were not performed. This consideration is important in order to clarify the important effort that the scientific community is still making to clarify all aspects related to COVID-19 vaccination.

Vaccines represent some of the greatest medical and scientific achievements of the modern era. In particular, in the pandemic scenario, a medico-legal perspective on vaccination is very important to provide a critical viewpoint. COVID-19 vaccine administration involves different questions, such as the possibility of side effects, which led to a fairly diffuse suspicion and rejection by many people, especially when a fatal case occurring at the same time as vaccine administration occurred in a young healthy subject. The use of vaccines also poses various ethical and legal problems, including the possibility of conflicts between individual and collective rights [34,35].

In a recent report [47], the link between vaccinations and the principles of biomedical ethics has been discussed by assessing four fundamental principles: autonomy (freedom of choice: mandatory and non-mandatory), non-maleficence (not causing harm), beneficence (promoting good), and legality. Several international authors have focused on this topic individually [48,49], suggesting that an international vaccination program should follow seven ethical principles [49]:

- The vaccination plan should concern a disease that represents a public health issue;
- The vaccine should be safe and effective;
- The distress to participants should be as low as possible;
- The benefit/risk ratio of the program must be favorable for participants;
- The immunization program should give the population an equal share of the benefits and burdens;
- The involvement should be, in general, voluntary, except where compulsory vaccination is essential to prevent a real risk;
- Public trust in vaccination programs should be respected and preserved.

In the case of vaccination against COVID-19, different governments decided to vaccinate healthcare personnel as a priority, playing a critical role in infection control in healthcare facilities. Similar decisions have been applied worldwide [50–52]. Therefore, considering that most deaths were recorded in the elderly and so-called "fragile" subjects, the vaccine administration was prioritized to these categories, widening the range of subjects involved. In the case of COVID-19 infection, there has been much debate on mandatory vaccination, although at present there is freedom of choice for everyone except for several categories such as health workers. If individual health implies self-determination, i.e., the right of everyone to decide whether and how to treat themselves (in the extreme, even not to treat themselves or let themselves die), in the case of collective health this right may be limited or weakened.

5. Key Recommendations

On the basis of the discussed data, we want to suggest several key recommendations to improve awareness of vaccine safety:

- All pathologists should publish autopsy reports in peer-reviewed journals or alternatively, deposit these reports in national/international databases maintained by pathologist societies; in this way, it will be possible to examine the causality relationship worldwide, analyzing other vaccines;

- All pathologists should apply the WHO algorithm to define the causality relationship between vaccination and adverse effects. Analyzing the data of this review, this important tool was usually not applied, although its use is strongly encouraged to define the causality of an adverse event following vaccination (AEFI) [53];
- The scientific community should consider the opportunity to create an international database with all data on adverse effects related to the COVID-19 vaccination that may be implemented and consulted by scientists worldwide.

6. Conclusions

In this context, the scientific community must work hard to reduce the growing hesitation to vaccinate among the general population following several cases of fatal adverse reactions. In many cases, the opposition, in the case of vaccines, is linked to pseudo-scientific reasons (not supported by evidence) or utilitarian reasons, therefore, it is not a question of conscientious objection. This raises a very delicate question: to what extent can an adult transfer the possible negative consequences of his or her choices to the entire community, as in the case of COVID-19 vaccination.

The great challenge for the scientific community in the fight against COVID-19 is represented by the success of a global vaccination campaign and, in this light, it is important to provide scientific evidence to remove the doubt of public opinion. In order to avoid another „Lockdown of science" [54,55], we are firmly convinced that autopsy should be the rule in the causality assessment of fatal cases occurring at the same time as COVID-19 vaccination. Measures such as clarifying vaccine safety and effectiveness are essential to reduce vaccine hesitancy in the general population. Despite vaccine hesitancy being a global phenomenon, the causes are very different in each country. As discussed in the present review, the reports about the severe adverse effects of vaccination, such as thrombosis, thrombocytopenia, and myocarditis, have negatively influenced public opinion, slowing down the vaccine program. In line with these considerations, it is desirable that all data collected after post-mortem investigations are shared in the scientific community in order to point out the relative countermeasures.

The development and large-scale implementation of COVID-19 vaccination represents a promising tool to achieve herd immunity, with the possibility to stop this global crisis. Many issues must be addressed regarding current approaches to vaccination to build an effective and correct public health response, building preparedness for future outbreaks.

Author Contributions: Conceptualization, F.S. and C.P.; methodology, F.S., M.S., M.E., N.D.N., P.Z. and C.P.; validation, F.S., M.S., M.E., N.D.N., P.Z. and C.P.; formal analysis, F.S., M.S. and C.P.; investigation, F.S., M.S., C.P.; data curation, F.S. and C.P.; writing—original draft preparation, F.S. and C.P.; writing—review and editing, F.S., M.S., M.E., N.D.N., P.Z. and C.P. All authors have read and agreed to the published version of the manuscript.

Funding: This research received no external funding.

Institutional Review Board Statement: Not applicable.

Informed Consent Statement: Informed consent is not required as long as information is anonymized and the submission does not include images that may identify the person.

Data Availability Statement: All data generated or analyzed during this study are included in this published article.

Acknowledgments: The authors thank the Scientific Bureau of the University of Catania for language support.

Conflicts of Interest: The authors declare no conflict of interest.

References

1. Gorbalenya, A.E.; Baker, S.C.; Baric, R.S.; de Groot, R.J.; Drosten, C.; Gulyaeva, A.A.; Haagmans, B.L.; Lauber, C.; Leontovich, A.M.; Neuman, B.W.; et al. The species Severe acute respiratory syndrome-related coronavirus: Classifying 2019-nCoV and naming it SARS-CoV-2. *Nat. Microbiol.* **2020**, *5*, 536–544.
2. Wordometers Info COVID-19 Coronavirus Pandemic—Wordometer. Available online: https://www.worldometers.info/coronavirus/ (accessed on 2 December 2021).
3. Josh Holder Tracking Coronavirus Vaccinations Around the World. Available online: https://www.nytimes.com/interactive/2021/world/COVID-vaccinations-tracker.html (accessed on 2 December 2021).
4. Mallapaty, S. India's DNA COVID vaccine is a world first—More are coming. *Nature* **2021**, *597*, 161–162. [CrossRef]
5. Edwards, K.; Orenstein, W. COVID-19: Vaccines to Prevent SARS-CoV-2 Infection. Available online: https://www.uptodate.com/contents/COVID-19-vaccines-to-prevent-SARS-CoV-2-infection (accessed on 2 December 2021).
6. Tregoning, J.S.; Flight, K.E.; Higham, S.L.; Wang, Z.; Pierce, B.F. Progress of the COVID-19 vaccine effort: Viruses, vaccines and variants versus efficacy, effectiveness and escape. *Nat. Rev. Immunol.* **2021**, *21*, 626–636. [CrossRef] [PubMed]
7. Haas, E.J.; Angulo, F.J.; McLaughlin, J.M.; Anis, E.; Singer, S.R.; Khan, F.; Brooks, N.; Smaja, M.; Mircus, G.; Pan, K.; et al. Impact and effectiveness of mRNA BNT162b2 vaccine against SARS-CoV-2 infections and COVID-19 cases, hospitalisations, and deaths following a nationwide vaccination campaign in Israel: An observational study using national surveillance data. *Lancet* **2021**, *397*, 1819–1829. [CrossRef]
8. Rzymski, P.; Pazgan-Simon, M.; Simon, K.; Łapiński, T.; Zarębska-Michaluk, D.; Szczepańska, B.; Chojnicki, M.; Mozer-Lisewska, I.; Flisiak, R. Clinical Characteristics of Hospitalized COVID-19 Patients Who Received at Least One Dose of COVID-19 Vaccine. *Vaccines* **2021**, *9*, 781. [CrossRef] [PubMed]
9. EMA. EMA Recommends First COVID-19 Vaccine for Authorisation in the EU. Available online: https://www.ema.europa.eu/en/news/ema-recommends-first-COVID-19-vaccine-authorisation-eu (accessed on 2 December 2021).
10. EMA. EMA Recommends COVID-19 Vaccine Moderna for Authorisation in the EU. Available online: https://www.ema.europa.eu/en/news/ema-recommends-COVID-19-vaccine-moderna-authorisation-eu (accessed on 2 December 2021).
11. EMA. EMA Recommends COVID-19 Vaccine AstraZeneca for Authorisation in the EU. Available online: https://www.ema.europa.eu/en/news/ema-recommends-COVID-19-vaccine-astrazeneca-authorisation-eu (accessed on 2 December 2021).
12. EMA. EMA Recommends COVID-19 Vaccine Janssen for Authorisation in the EU. Available online: https://www.ema.europa.eu/en/news/ema-recommends-COVID-19-vaccine-janssen-authorisation-eu (accessed on 2 December 2021).
13. EMA. COVID-19 Vaccines. Available online: https://www.ema.europa.eu/en/human-regulatory/overview/public-health-threats/coronavirus-disease-COVID-19/treatments-vaccines/COVID-19-vaccines (accessed on 2 December 2021).
14. EMA Pharmacovigilance: Overview. Available online: https://www.ema.europa.eu/en/human-regulatory/overview/pharmacovigilance-overview (accessed on 2 December 2021).
15. European Centre for Disease Prevention and Control. *Suspected Adverse Reactions to COVID-19 Vaccination and the Safety of Substances of Human Origin*; ECDC: Stockholm, Sweden, 2021.
16. Schwarzinger, M.; Watson, V.; Arwidson, P.; Alla, F.; Luchini, S. COVID-19 vaccine hesitancy in a representative working-age population in France: A survey experiment based on vaccine characteristics. *Lancet Public Health* **2021**, *6*, e210–e221. [CrossRef]
17. Solís Arce, J.S.; Warren, S.S.; Meriggi, N.F.; Scacco, A.; McMurry, N.; Voors, M.; Syunyaev, G.; Malik, A.A.; Aboutajdine, S.; Adeojo, O.; et al. COVID-19 vaccine acceptance and hesitancy in low- and middle-income countries. *Nat. Med.* **2021**, *27*, 1385–1394. [CrossRef] [PubMed]
18. Moher, D.; Shamseer, L.; Clarke, M.; Ghersi, D.; Liberati, A.; Petticrew, M.; Shekelle, P.; Stewart, L.A. Preferred reporting items for systematic review and meta-analysis protocols (PRISMA-P) 2015 statement. *Syst. Rev.* **2015**, *4*, 1. [CrossRef] [PubMed]
19. Greinacher, A.; Thiele, T.; Warkentin, T.E.; Weisser, K.; Kyrle, P.A.; Eichinger, S. Thrombotic Thrombocytopenia after ChAdOx1 nCov-19 Vaccination. *N. Engl. J. Med.* **2021**, *384*, 2092–2101. [CrossRef]
20. Althaus, K.; Möller, P.; Uzun, G.; Singh, A.; Beck, A.; Bettag, M.; Bösmüller, H.; Guthoff, M.; Dorn, F.; Petzold, G.C.; et al. Antibody-mediated procoagulant platelets in SARS-CoV-2- vaccination associated immune thrombotic thrombocytopenia. *Haematologica* **2021**, *106*, 2170. [CrossRef]
21. Mauriello, A.; Scimeca, M.; Amelio, I.; Massoud, R.; Novelli, A.; Di Lorenzo, F.; Finocchiaro, S.; Cimino, C.; Telesca, R.; Chiocchi, M.; et al. Thromboembolism after COVID-19 vaccine in patients with preexisting thrombocytopenia. *Cell Death Dis.* **2021**, *12*, 762. [CrossRef]
22. Wiedmann, M.; Skattør, T.; Stray-Pedersen, A.; Romundstad, L.; Antal, E.-A.; Marthinsen, P.B.; Sørvoll, I.H.; Leiknes Ernstsen, S.; Lund, C.G.; Holme, P.A.; et al. Vaccine Induced Immune Thrombotic Thrombocytopenia Causing a Severe Form of Cerebral Venous Thrombosis With High Fatality Rate: A Case Series. *Front. Neurol.* **2021**, *12*, 721146. [CrossRef]
23. Bjørnstad-Tuveng, T.H.; Rudjord, A.; Anker, P. Fatal cerebral haemorrhage after COVID-19 vaccine. *Tidsskr. Nor. Laegeforen.* **2021**, *141*. [CrossRef]
24. Scully, M.; Singh, D.; Lown, R.; Poles, A.; Solomon, T.; Levi, M.; Goldblatt, D.; Kotoucek, P.; Thomas, W.; Lester, W. Pathologic Antibodies to Platelet Factor 4 after ChAdOx1 nCoV-19 Vaccination. *N. Engl. J. Med.* **2021**, *384*, 2202–2211. [CrossRef]
25. Günther, A.; Brämer, D.; Pletz, M.W.; Kamradt, T.; Baumgart, S.; Mayer, T.E.; Baier, M.; Autsch, A.; Mawrin, C.; Schönborn, L.; et al. Complicated Long Term Vaccine Induced Thrombotic Immune Thrombocytopenia-A Case Report. *Vaccines* **2021**, *9*, 1344. [CrossRef]

26. Pomara, C.; Sessa, F.; Ciaccio, M.; Dieli, F.; Esposito, M.; Garozzo, S.F.; Giarratano, A.; Prati, D.; Rappa, F.; Salerno, M.; et al. Post-mortem findings in vaccine-induced thrombotic thombocytopenia. *Haematologica* **2021**, *106*, 2291. [CrossRef]
27. Pomara, C.; Sessa, F.; Ciaccio, M.; Dieli, F.; Esposito, M.; Giammanco, G.M.; Garozzo, S.F.; Giarratano, A.; Prati, D.; Rappa, F.; et al. COVID-19 Vaccine and Death: Causality Algorithm According to the WHO Eligibility Diagnosis. *Diagnostics* **2021**, *11*, 955. [CrossRef] [PubMed]
28. Schneider, J.; Sottmann, L.; Greinacher, A.; Hagen, M.; Kasper, H.-U.; Kuhnen, C.; Schlepper, S.; Schmidt, S.; Schulz, R.; Thiele, T.; et al. Postmortem investigation of fatalities following vaccination with COVID-19 vaccines. *Int. J. Legal Med.* **2021**, *135*, 2335–2345. [CrossRef]
29. Edler, C.; Klein, A.; Schröder, A.S.; Sperhake, J.-P.; Ondruschka, B. Deaths associated with newly launched SARS-CoV-2 vaccination (Comirnaty®). *Leg. Med.* **2021**, *51*, 101895. [CrossRef] [PubMed]
30. Hansen, T.; Titze, U.; Kulamadayil-Heidenreich, N.S.A.; Glombitza, S.; Tebbe, J.J.; Röcken, C.; Schulz, B.; Weise, M.; Wilkens, L. First case of postmortem study in a patient vaccinated against SARS-CoV-2. *Int. J. Infect. Dis.* **2021**, *107*, 172–175. [CrossRef] [PubMed]
31. Choi, S.; Lee, S.; Seo, J.-W.; Kim, M.-J.; Jeon, Y.H.; Park, J.H.; Lee, J.K.; Yeo, N.S. Myocarditis-induced Sudden Death after BNT162b2 mRNA COVID-19 Vaccination in Korea: Case Report Focusing on Histopathological Findings. *J. Korean Med. Sci.* **2021**, *36*, e286. [CrossRef] [PubMed]
32. Verma, A.K.; Lavine, K.J.; Lin, C.-Y. Myocarditis after COVID-19 mRNA Vaccination. *N. Engl. J. Med.* **2021**, *385*, 1332–1334. [CrossRef] [PubMed]
33. Choi, G.-J.; Baek, S.H.; Kim, J.; Kim, J.H.; Kwon, G.-Y.; Kim, D.K.; Jung, Y.H.; Kim, S. Fatal Systemic Capillary Leak Syndrome after SARS-CoV-2Vaccination in Patient with Multiple Myeloma. *Emerg. Infect. Dis.* **2021**, *27*, 2973–2975. [CrossRef] [PubMed]
34. Maltezou, H.C.; Theodoridou, K.; Ledda, C.; Rapisarda, V.; Theodoridou, M. Vaccination of healthcare workers: Is mandatory vaccination needed? *Expert Rev. Vaccines* **2019**, *18*, 5–13. [CrossRef]
35. Scavone, C.; Sessa, M.; Clementi, E.; Rossi, F.; Capuano, A. Italian Immunization Goals: A Political or Scientific Heated Debate? *Front. Pharmacol.* **2018**, *9*, 574. [CrossRef] [PubMed]
36. Sharifian-Dorche, M.; Bahmanyar, M.; Sharifian-Dorche, A.; Mohammadi, P.; Nomovi, M.; Mowla, A. Vaccine-induced immune thrombotic thrombocytopenia and cerebral venous sinus thrombosis post COVID-19 vaccination; a systematic review. *J. Neurol. Sci.* **2021**, *428*, 117607. [CrossRef]
37. Arepally, G.M.; Ortel, T.L. Vaccine-induced immune thrombotic thrombocytopenia: What we know and do not know. *Blood* **2021**, *138*, 293–298. [CrossRef] [PubMed]
38. Foddy, B.; Savulescu, J. Addiction and autonomy: Can addicted people consent to the prescription of their drug of addiction? *Bioethics* **2006**, *20*, 1–15. [CrossRef]
39. Salerno, M.; Mizio, G.D.; Montana, A.; Pomara, C. To be or not to be vaccinated? That is the question among Italian healthcare workers: A medico-legal perspective. *Future Microbiol.* **2019**, *14*, 51–54. [CrossRef]
40. Mungmunpuntipamtip, R.; Wiwanitkit, V. Deaths associated with newly launched SARS-CoV-2 vaccination. *Leg. Med.* **2021**, *53*, 101956. [CrossRef]
41. Sessa, F.; Salerno, M.; Pomara, C. Autopsy Tool in Unknown Diseases: The Experience with Coronaviruses (SARS-CoV, MERS-CoV, SARS-CoV-2). *Med.* **2021**, *57*, 309. [CrossRef]
42. Pomara, C.; Volti, G.L.; Cappello, F. COVID-19 Deaths: Are We Sure It Is Pneumonia? Please, Autopsy, Autopsy, Autopsy! *J. Clin. Med.* **2020**, *9*, 1259. [CrossRef] [PubMed]
43. Rzymski, P.; Perek, B.; Flisiak, R. Thrombotic Thrombocytopenia after COVID-19 Vaccination: In Search of the Underlying Mechanism. *Vaccines* **2021**, *9*, 559. [CrossRef]
44. Shastri, J.; Parikh, S.; Aggarwal, V.; Agrawal, S.; Chatterjee, N.; Shah, R.; Devi, P.; Mehta, P.; Pandey, R. Severe SARS-CoV-2 Breakthrough Reinfection With Delta Variant After Recovery From Breakthrough Infection by Alpha Variant in a Fully Vaccinated Health Worker. *Front. Med.* **2021**, *8*, 1379. [CrossRef] [PubMed]
45. Ledford, H. Coronavirus reinfections: Three questions scientists are asking. *Nature* **2020**, *585*, 168–169. [CrossRef]
46. Mevorach, D.; Anis, E.; Cedar, N.; Bromberg, M.; Haas, E.J.; Nadir, E.; Olsha-Castell, S.; Arad, D.; Hasin, T.; Levi, N.; et al. Myocarditis after BNT162b2 mRNA Vaccine against COVID-19 in Israel. *N. Engl. J. Med.* **2021**, *385*, 2140–2149. [CrossRef]
47. Van Rostenberghe, H. Primum Non Nocere. *Malays. J. Med. Sci.* **2021**, *28*, 122–124. [CrossRef]
48. Salmon, D.A.; Haber, M.; Gangarosa, E.J.; Phillips, L.; Smith, N.J.; Chen, R.T. Health Consequences of Religious and Philosophical Exemptions From Immunization LawsIndividual and Societal Risk of Measles. *JAMA* **1999**, *282*, 47–53. [CrossRef] [PubMed]
49. Verweij, M.; Dawson, A. Ethical principles for collective immunisation programmes. *Vaccine* **2004**, *22*, 3122–3126. [CrossRef]
50. Rimmer, A. COVID-19: Government considers mandatory vaccination for healthcare staff in England. *BMJ* **2021**, *374*, n2222. [CrossRef]
51. Paterlini, M. COVID-19: Italy makes vaccination mandatory for healthcare workers. *BMJ* **2021**, *373*, n905. [CrossRef] [PubMed]
52. Chew, N.W.S.; Cheong, C.; Kong, G.; Phua, K.; Ngiam, J.N.; Tan, B.Y.Q.; Wang, B.; Hao, F.; Tan, W.; Han, X.; et al. An Asia-Pacific study on healthcare workers' perceptions of, and willingness to receive, the COVID-19 vaccination. *Int. J. Infect. Dis.* **2021**, *106*, 52–60. [CrossRef] [PubMed]
53. WHO. *Causality Assessment Of An Adverse Event Following Immunization*, 2nd ed.; WHO: Geneva, Switzerland, 2019; ISBN 9789241513654.

54. Salerno, M.; Sessa, F.; Piscopo, A.; Montana, A.; Torrisi, M.; Patanè, F.; Murabito, P.; Li Volti, G.; Pomara, C. No Autopsies on COVID-19 Deaths: A Missed Opportunity and the Lockdown of Science. *J. Clin. Med.* **2020**, *9*, 1472. [CrossRef]
55. Sessa, F.; Bertozzi, G.; Cipolloni, L.; Baldari, B.; Cantatore, S.; Errico, S.D.; Mizio, G.D.; Asmundo, A.; Castorina, S.; Salerno, M.; et al. Clinical-Forensic Autopsy Findings to Defeat COVID-19 Disease: A Literature Review. *J. Clin. Med.* **2020**, *9*, 2026. [CrossRef] [PubMed]

Journal of
Clinical Medicine

Article

Determining What Changed Japanese Suicide Mortality in 2020 Using Governmental Database

Ryusuke Matsumoto, Eishi Motomura, Kouji Fukuyama, Takashi Shiroyama and Motohiro Okada *

Department of Neuropsychiatry, Division of Neuroscience, Graduate School of Medicine, Mie University, Tsu 514-8507, Japan; matsumoto-r@clin.medic.mie-u.ac.jp (R.M.); motomura@clin.medic.mie-u.ac.jp (E.M.); k-fukuyama@clin.medic.mie-u.ac.jp (K.F.); takashi@clin.medic.mie-u.ac.jp (T.S.)
* Correspondence: okadamot@clin.medic.mie-u.ac.jp; Tel.: +81-59-231-5018

Abstract: The pandemic of 2019 novel coronavirus disease (COVID-19) caused both COVID-19-related health hazards and the deterioration of socioeconomic and sociopsychological status due to governmental restrictions. There were concerns that suicide mortality would increase during the COVID-19 pandemic; however, a recent study reported that suicide mortality did not increase in 21 countries during the early pandemic period. In Japan, suicide mortality was reduced from 2009 to 2019, but both the annual number of suicide victims and the national suicide mortality rates in 2020 increased compared to that in 2019. To clarify the discrepancy of suicide mortality between the first and second half of 2020 in Japan, the present study determines annual and monthly suicide mortality disaggregated by prefectures, gender, age, means, motive, and household factors during the COVID-19 pandemic and pre-pandemic periods using a linear mixed-effects model. Furthermore, the relationship between suicide mortality and COVID-19 data (the infection rate, mortality, and duration of the pandemic) was analysed using hierarchal linear regression with a robust standard error. The average of monthly suicide mortality of both males and females in all 47 prefectures decreased during the first stay-home order (April–May) (females: from 10.1–10.2 to 7.8–7.9; males: from 24.0–24.9 to 21.6 per 100,000 people), but increased after the end of the first stay-home order (July–December) (females: from 7.5–9.5 to 10.3–14.5; males: from 19.9–23.0 to 21.1–26.7 per 100,000 people). Increasing COVID-19-infected patients and victims indicated a tendency of suppression, but the prolongation of the pandemic indicated a tendency of increasing female suicide mortality without affecting that of males. Contrary to the national pattern, in metropolitan regions, decreasing suicide mortality during the first stay-home order was not observed. Decreasing suicide mortality during the first stay-home order was not observed in populations younger than 30 years old, whereas increasing suicide mortality of populations younger than 30 years old after the end of the first stay-home order was predominant. A decrease in suicide mortality of one-person household residents during the first stay-home order was not observed. The hanging suicide mortality of males and females was decreased and increased during and after the end of the first stay-home orders, respectively; however, there was no decrease in metropolitan regions. These results suggest that the suicide mortality in 2020 of females, younger populations, urban residents, and one-person household residents increased compared to those of males, the elderly, rural residents, and multiple-person household residents. Therefore, the unexpected drastic fluctuations of suicide mortality during the COVID-19 pandemic in Japan were probably composed of complicated reasons among various identified factors in this study, and other unknown factors.

Keywords: suicide mortality; Japan; COVID-19; gender; region; age; motive; means; household

Citation: Matsumoto, R.; Motomura, E.; Fukuyama, K.; Shiroyama, T.; Okada, M. Determining What Changed Japanese Suicide Mortality in 2020 Using Governmental Database. *J. Clin. Med.* **2021**, *10*, 5199. https://doi.org/10.3390/jcm10215199

Academic Editor: Giorgio I. Russo

Received: 20 September 2021
Accepted: 5 November 2021
Published: 7 November 2021

Publisher's Note: MDPI stays neutral with regard to jurisdictional claims in published maps and institutional affiliations.

1. Introduction

The 2019 novel coronavirus disease (COVID-19) had globally infected more than 242 million people and contributed to over 4.9 million deaths as of 25 October 2021 [1]. In Japan, more than 1.7 million people were infected with COVID-19, and more than

18,000 lives were lost due to COVID-19 as of 25 October 2021 [1,2]. Even today, when vaccines are more widespread, the COVID-19 pandemic continues. In addition to COVID-19-related health hazards, governmental COVID-19 restrictions have impacted lives and lifestyles, resulting in the deterioration of socioeconomic and sociopsychological status. A number of studies expressed sociopsychological concerns that the COVID-19 pandemic has encouraged isolation, fear, marginalisation, psychiatric disorders, domestic abuse, and intimate partner violence [3–7]. Social conditions that force the drastic modification of lifestyles and the economy play important roles in increasing suicide mortality [8,9]. Contrary to our expectations, a recent study revealed that the risk of suicide of 21 countries could not be detected during the early COVID-19 pandemic periods, including Japan [10]. However, an extended observation period reported that the prolongation of the COVID-19 pandemic periods increased suicide mortality in some countries and areas, such as Japan, Puerto Rico, and Vienna (Austria) [10].

Several analytical studies that have utilised governmental suicide databases covering the entire population reported that the impact of the COVID-19 pandemic on suicide mortality in Japan may change over time and could have various different targets [11,12]. In Japan, suicide mortality was steadily decreasing between 2009 and 2019, but increasing in 2020: from 20,169 (males: 14,078, females: 6091) to 20,919 (males: 13,943, females: 6976) (Figure 1) [13]. Analysis of gender-related dynamics of suicide mortality indicated confusing results [14–19]. The annual suicide mortality of males decreased compared to that in 2019, but that of females increased [15–19] (Figure 1). In various Asian countries including Japan, where an adverse effect of the economic crisis on suicide mortality was detected around the 2008 Asian economic crisis, the impact of the economic crisis on the suicide mortality of males and elderly populations was greater compared to that of females and younger populations (Figure 1A) [8,20]. According to this evidence, comprehensive suicide prevention programmes in Japan reduced suicide mortality due to targeting males and elderly populations [21–25]; however, the economic recession and increasing suicide mortality in Japan were not temporally related between 2009 and 2019 (Figure 1A). Therefore, increasing suicide mortality in 2020 was uninterpretable using previous findings. The Ministry of Health, Labour, and Welfare (MHLW) speculated that the increasing female suicide mortality was probably induced by mass media reports associated with the suicide and death of celebrities with COVID-19, and increasing domestic violence, as evidenced by the WHO guideline [26]. MHLW announced nine alerts in the mass media [27,28] from September 2020 to September 2021, issuing a warning that reporting should be conducted according to suicide reporting guidelines "Preventing suicide: a resource for media professionals" [26].

The first modern global pandemic was the influenza pandemic (Spanish flu) between 1918 and 1920. As well as the Spanish flu [29,30], other infectious pandemics whose impacts on suicide mortality have been analysed include severe acute respiratory syndrome (SARS) [31–33] and COVID-19. Regarding the Spanish flu pandemic, suicide mortality increased after the first pandemic phase in 1919, but an increase was not observed after the second pandemic phase [29]; however, details of the impacts of the Spanish flu pandemic are not clear due to suppressed reporting of the pandemic during World War I [30]. Regarding the SARS pandemic in Hong Kong, a persistently increasing suicide mortality rate (over 2 years) compared to the pre-pandemic period was detected [31–33]. Increasing and decreasing (valley of suicide) suicide mortality rates were detected that were synchronised with the peak of the SARS pandemic, as well as 2 months after the peak, but these fluctuations were limited in older females [31,32]. Notably, feeling disconnected was a more common problem in individuals who were identified as having committed suicide in relation to SARS than in those who were identified as having committed suicide for other reasons [33,34]. However, there is little information and evidence regarding the time-dependent changes in suicide mortality induced by the COVID-19 pandemic in Japan. Thus, a number of reports speculated on the basis of evidence relating to previous public health emergencies arising from natural disasters [7,34–38]. Short-term decreasing suicide rates in the immediate aftermath of natural disasters are called the "honeymoon period" or "pulling together" phenomenon [37,38]. Regional panel data analysis in Japan revealed that, when damage caused by natural disasters is extremely severe, suicide mortality tends to increase in the immediate aftermath of the disaster and several years later; however, when the damage by natural disasters is less severe, suicide mortality rates tend to decrease after the disasters, especially one or two years later [36]. Taken together with previous findings, the time-dependent kinetics of Japanese suicide mortality in 2020 is more similar to the tendency induced by less severe natural disasters rather than that by severe natural disasters or an economic crisis. The COVID-19 pandemic has imposed a slow and long-term burden on society and individuals compared to severe natural disasters such as earthquakes and tsunamis. Indeed, increasing suicide mortality rates are concerning due to the prolongation of the COVID-19 pandemic, since the rate of suicidal ideations during the COVID-19 pandemic is higher than that reported in studies on the general population during the pre-pandemic period [39]. Therefore, the long-lasting burden on societies and individuals induced by the COVID-19 pandemic is fundamentally different from that of previous natural disasters and economic crises. John et al. reported that "any change in the risk of suicide associated with COVID-19 is likely to be dynamic" [40]. Therefore, to clarify who, when, how, and why individuals died by suicide in Japan during the COVID-19 pandemic, this study determines the change in suicide mortality during the pandemic compared to the pre-pandemic period using a governmental database.

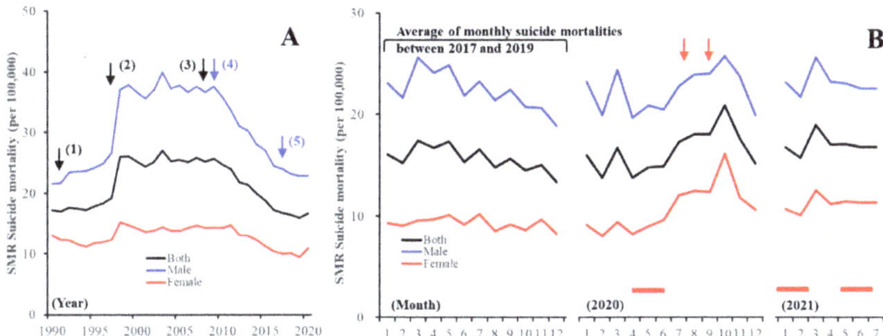

Figure 1. (**A**) Annual and (**B**) monthly suicide mortality in Japan. Ordinates indicate standardised mortality ratio (SMR) of suicide mortality of males and females (black lines), males (blue line), and females (red line) in Japan per 100,000 people. (**A**) Black arrows (1–3) indicate the collapse of the asset bubble, the Asian economic crisis, and the 2008 global financial crisis, respectively. Blue arrows (4) and (5) indicate the contribution of the Emergency Fund to Enhance Community-Based Suicide Countermeasures and the introduction of the Revised Basic Act on Suicide Prevention, respectively. (**B**) (left) Average SMR suicide mortalities during 2017–2019. (middle, right) Monthly SMR suicide mortality in 2020 and 2021, respectively. Red columns, period of 2019 novel coronavirus disease (COVID-19) stay-home orders; red arrows, mass media reports regarding celebrity suicides.

2. Materials and Methods

2.1. Dependent and Independent Variables

The numbers of suicide victims of 47 prefectures in Japan from January 2017 to June 2021 were obtained from Basic Data on Suicide in the Region (BDSR) in a national database of the MHLW [13]. BDSR published the numbers of suicide victims disaggregated by prefecture (47 prefectures), gender, suicide motives (health, family, economy, romance, employment, and school-related motives), suicide means (hanging, charcoal burning, jumping, poisoning, and throwing), household (multiple-person and one-person), and ages (0–19 (10s), 20–29 (20s), 30–39 (30s), 40–49 (40s), 50–59 (50s), 60–69 (60s), 70–79 (70s), and over 80 years old (80s)) [13]. The numbers of suicide victims in BDSR between January 2017 and December 2020 were definitive values, but as of September 2021, suicide victims between January and June 2021 are provisional values (final definitive value will be published in March 2022). Prefectural population data were obtained from the Regional Statistics Database (RSD) of the System of Social and Demographic Statistics of the Statistics Bureau of the Ministry of Internal Affairs and Communications (SBMIAC) [41].

Annual standardised mortality of suicide per 100,000 people (SMR-S) was calculated by dividing the numbers of suicide victims per prefecture by the prefectural population (denominator) of the same years. Monthly SMR-S, which was also calculated by dividing monthly numbers of suicide victims per prefecture by the prefectural population of the same years, was converted annually and adopted for statistical analysis. Annual and monthly age, gender, and prefecture disaggregated SMR-S were also derived from age, gender, and prefecture disaggregated numbers of suicide victims (numerator) in BDSR [13], and age, gender, and prefecture disaggregated population exposure (denominator) [41] in RDS [22]. BDSR data were classified by the number of suicides into six types of suicide motives (health-, family-, economy-, romance-, employment-, and school-related problems), five types of suicidal means (hanging, poisoning, charcoal burning, jumping, and throwing), and household conditions (one- and multiple-person households) [13]. Suicide victims in each region were counted by the jurisdiction of local police stations. The police investigate personal characteristics and background factors of each suicide victim. The results of this investigation contain a number of motives for suicide, and these motives were compared to previously compiled suicide motive lists. Lastly, the investigation identifies the possible motive for suicide on the basis of evidence, suicide notes, or other documentation such as medical certificates, clinical recording, and the testimony of the surviving family [42,43].

The monthly numbers of infected individuals with COVID-19 and death caused by COVID-19 were obtained from the Database of the National Institute of Infectious Diseases [2] and Sapporo Medical University School of Medicine [44]. The monthly COVID-19 infection ratio per 100,000 people (SCR) and the mortality ratio caused by COVID-19 per 100,000 people (SMR-C) were calculated by dividing the numbers of infected individuals and deaths caused by COVID-19 per prefecture by the prefectural population (denominator) of the same years. The monthly duration of the COVID-19 pandemic (DCP) was set as the monthly basis, with March 2020 as one month, since the SCR was drastically increased in March 2020 (first Japanese patients with COVID-19 and victims who had never travelled to China were confirmed in January and February 2020, respectively).

2.2. Statistical Analysis

The annual SMR-S between the pre-pandemic (between 2017 and 2019) and COVID-19 pandemic (between 2020 and 2021) periods was compared using a linear mixed-effects model using BellCurve for Excel v.3.2 (Social Survey Research Information Co., Ltd., Tokyo, Japan) [45–47]. Monthly SMR-S between the COVID-19 pandemic period (between January 2020 and June 2021) and the average of SMR-S at the same month pre-pandemic (between January 2017 and December 2019) were also compared by linear mixed-effects model using BellCurve for Excel v.3.2. When the F value of the linear mixed-effects model was significant ($p < 0.05$), data were analysed by Tukey's multiple comparison test. The governmental guideline for suicide prevention in Japan was stipulated in the General Policies for Comprehensive Measures against Suicide, which was revised in 2017. This revised guideline required prefectures to majorly improve scientific evidence-based regional suicide prevention programmes. Therefore, the pre-pandemic period was set between 2017 and 2019. The present study analysed the impact of SCR (monthly standardised COVID-19 infection ratio), SMR-C (monthly standardised mortality ratio caused by COVID-19), and DCP (duration of COVID-19 pandemic) on monthly SMR-S (between March 2020 and June 2021) using a hierarchical linear regression model with robust standard error (HLM7, Scientific Software International, Skokie, IL, USA) [23,24].

First, both linear mixed-effects and hierarchical linear regression models were analysed in all 47 prefectures. The target regions of the first governmental stay-home order (between April and May 2020) were major metropolitan regions in Japan, such as Tokyo, Saitama, Chiba, and Kanagawa, the Kansai area (Osaka, Kyoto, and Hyogo, Japan), the

Chukyo area (Aichi), the Fukuoka area (Fukuoka), and the Sapporo area (Hokkaido, Japan). Second, both linear mixed-effects and hierarchical linear regression models were analysed in metropolitan regions as serious COVID-19 infection areas.

3. Results

3.1. Suicide Mortality Disaggregated by Gender

Linear mixed-effect models detected the significantly different annual and monthly SMR-S of all 47 prefectures between COVID-19 pre-pandemic (between 2017 and 2019) and pandemic (between 2020 and 2021) periods. In all 47 prefectures, the annual SMR-S of females was increased in 2020, but that of males was decreased compared to average SMR-S during the pre-pandemic period (Figures 2 and 3). In 2020, the monthly SMR-S of males and females decreased during the first stay-home order (between April and May 2020) compared to the average SMR-S at the same month during the pre-pandemic period (Table 1, Figures 2 and 3). The monthly SMR-S of both males and females increased between the first and second stay-home orders (between August and December 2020) compared to the average SMR-S in the same month during the pre-pandemic period (Table 1, Figures 2 and 3). In 2021, the trends of the monthly SMR-S of males and females were different from those in 2020. The SMR-S of both males and females during the second stay-home order (between January and March 2021) increased, but there were no differences in that of both males and females during the third stay-home order (between May and July 2021) compared to the pre-pandemic period (Table 1, Figures 2 and 3).

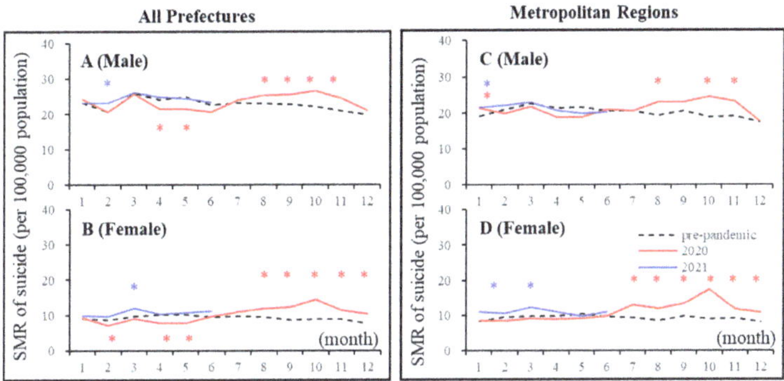

Figure 2. Temporal fluctuations of monthly standardised suicide mortalities (SMR-S) during COVID-19 pandemic period (2020–2021) compared to the average of the same month of SMR-S during the pre-pandemic period (2017–2019). Dotted black, red, and blue lines indicate the average of SMR-S for (**A**) males, (**B**) females of all 47 prefectures, and (**C**) males and (**D**) females of metropolitan regions. Ordinates indicate the SMR-S (per 100,000 people), and abscissas indicate the month. * $p < 0.05$, significant change using a linear mixed-effects model with Tukey's multiple comparison. Red and blue asterisks indicate significant changes in SMR-S in 2020 and 2021, respectively, compared to the average SMR-S of the same month during the pre-pandemic period.

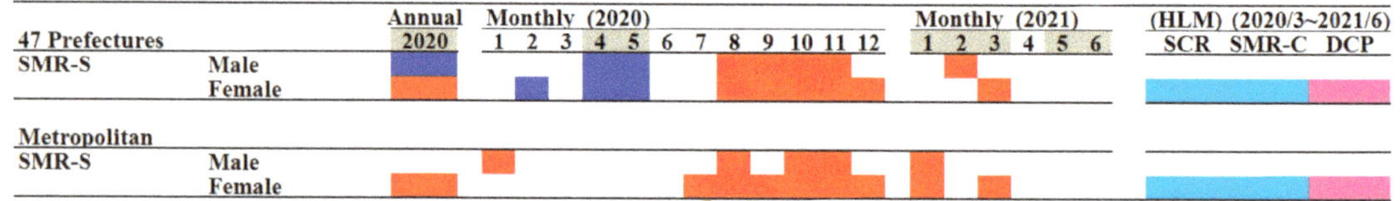

Figure 3. Comparison of annual and monthly standardised suicide mortalities (SMR-S) during COVID-19 pandemic period compared to the average of the same month of SMR-S during the pre-pandemic period (2017–2019), and the impact of the COVID-19 pandemic on SMR-S. Blue and red columns indicate significant decreasing and increasing suicide mortalities using a linear mixed-effects model with Tukey's multiple comparison ($p < 0.05$). Light blue and red columns indicate significant decrease and increase factors against SMR-S using hierarchical linear model with robust standard error ($p < 0.05$). SMR-S: standardised suicide mortality per 100,000 people. SCR: standardised infection with COVID-19 per 100,000 people. SMR-C: standardised mortality caused by COVID-19 per 100,000 people. DCP: during COVID-19 pandemic.

Table 1. Monthly standardised suicide mortalities (SMR-S) of males and females in all 47 prefectures and five metropolitan regions (per 100,000 people).

			All 47 Prefectures								Five Metropolitan Regions							
		Male				Female				Male				Female			
Year	Month	Mean	±	SD	(p)	Mean	±	SD	(p)	Mean	±	SD	(p)	Mean	±	SD	(p)
Pre-pandemic Average (2017–2019)	1	23.2	±	4.8		9.1	±	2.2		19.2	±	3.3		8.4	±	1.4	
	2	20.6	±	4.3		8.6	±	2.3		21.0	±	3.3		9.6	±	1.4	
	3	26.1	±	4.7		9.7	±	2.8		22.6	±	2.9		9.8	±	1.5	
	4	24.0	±	4.5		10.2	±	2.7		21.4	±	2.6		9.9	±	1.5	
	5	24.9	±	5.0		10.1	±	2.8		21.7	±	4.0		10.5	±	1.0	
	6	22.7	±	4.7		9.5	±	2.1		20.5	±	2.2		9.5	±	1.4	
	7	23.2	±	5.3		9.8	±	2.6		20.7	±	3.0		9.4	±	1.0	
	8	23.0	±	4.7		9.5	±	2.6		19.4	±	2.7		8.6	±	1.5	
	9	22.8	±	5.1		8.6	±	2.2		20.5	±	3.7		9.7	±	1.3	
	10	22.1	±	4.9		8.8	±	2.1		19.0	±	2.5		9.0	±	1.3	
	11	21.0	±	4.5		9.0	±	2.1		19.2	±	3.6		9.2	±	1.2	
	12	19.9	±	3.4		7.5	±	2.0		17.4	±	2.6		8.1	±	1.2	
2020	1	24.3	±	6.1	(0.35)	9.3	±	3.5	(0.75)	21.3	±	3.4	(0.03) *	8.4	±	1.7	(0.93)
	2	20.8	±	6.4	(0.88)	7.3	±	3.1	(0.01) **	20.1	±	3.0	(0.32)	8.6	±	2.8	(0.30)
	3	25.9	±	7.8	(0.83)	9.0	±	3.9	(0.27)	21.8	±	2.6	(0.47)	9.2	±	1.8	(0.46)
	4	21.6	±	6.1	(0.01) *	7.8	±	3.9	(0.00) **	19.0	±	3.5	(0.08)	8.9	±	1.7	(0.22)
	5	21.6	±	8.4	(0.01) **	7.9	±	3.9	(0.00) **	18.9	±	3.4	(0.11)	9.0	±	2.5	(0.08)
	6	20.7	±	5.9	(0.08)	9.7	±	3.8	(0.76)	21.1	±	3.4	(0.51)	9.7	±	1.8	(0.81)
	7	24.2	±	7.3	(0.32)	10.9	±	4.7	(0.09)	20.5	±	3.4	(0.93)	12.9	±	3.4	(0.01) **
	8	25.4	±	7.1	(0.01) *	11.7	±	4.9	(0.00) **	23.0	±	2.1	(0.00) **	11.9	±	3.2	(0.01) *
	9	25.7	±	6.1	(0.02) *	12.3	±	4.3	(0.00) **	23.0	±	3.0	(0.07)	13.4	±	1.9	(0.00) **
	10	26.7	±	6.9	(0.00) **	14.5	±	5.5	(0.00) **	24.4	±	4.4	(0.00) **	17.3	±	1.8	(0.00) **
	11	24.6	±	5.8	(0.00) **	11.4	±	4.0	(0.00) **	23.2	±	2.8	(0.01) **	11.7	±	2.8	(0.02) *
	12	21.1	±	6.3	(0.18)	10.3	±	4.6	(0.00) **	17.6	±	1.4	(0.82)	10.8	±	2.3	(0.03) *
2021	1	23.3	±	6.5	(1.00)	10.0	±	3.8	(0.37)	21.6	±	3.8	(0.03) *	11.0	±	1.6	(0.00) **
	2	23.2	±	5.5	(0.03) *	9.7	±	4.3	(0.19)	22.2	±	4.2	(0.54)	10.6	±	1.9	(0.48)
	3	26.3	±	8.3	(0.99)	12.0	±	4.6	(0.00) **	23.1	±	3.8	(0.93)	12.4	±	2.9	(0.03) *
	4	24.9	±	8.5	(0.71)	10.4	±	3.4	(0.97)	20.7	±	6.1	(0.89)	11.1	±	2.3	(0.22)
	5	24.5	±	8.2	(0.94)	10.7	±	3.5	(0.52)	19.9	±	6.4	(0.68)	9.7	±	2.7	(0.66)
	6	23.5	±	7.7	(0.80)	11.2	±	4.8	(0.06)	20.3	±	3.6	(0.99)	11.1	±	3.1	(0.25)

* $p < 0.05$, ** $p < 0.05$: significant change compared to average SMR-S of the same month during pre-pandemic period (between 2017 and 2019) using a linear mixed-effects model with Tukey's multiple comparison. (p), *p* values of linear mixed-effects model with Tukey's multiple comparison.

These results indicate that significant fluctuations were observed in the suicide mortality of both males and females in 2020, which were classified into two phases: decreasing suicide during the first stay-home order and increasing suicide between the first and second stay-home orders. The suicide mortality of males and females in 2021 is expected to be slightly higher than that in pre-pandemic periods, but it seems to be stabilising, at least compared to the fluctuations of suicide mortality in 2020. The response of suicide mortality to the stay-home order was probably attenuated in a frequency-dependent manner of announcements of the stay-at-home orders.

In five major metropolitan regions that were the most severe pandemic regions (capital, Kansai, Chukyo, Fukuoka, and Sapporo metropolitan regions), the linear mixed-effect model detected significantly different annual and monthly SMR-S between the pre-pandemic (between 2017 and 2019) and pandemic periods (between 2020 and 2021) (Table 1, Figures 2 and 3). The annual SMR-S of females in 2020 increased, but that of males did not change compared to the average SMR-S of five metropolitan regions during the pre-pandemic period (Table 1, Figures 2 and 3). In 2020, the monthly SMR-S of males and females did not change during the first stay-home order compared to the average SMR-S during the pre-pandemic period; however, similar to all 47 prefectures, the monthly SMR-S of males and females increased between the first and second stay-home orders compared to the average SMR-S during the pre-pandemic period (Table 1, Figures 2 and 3). In 2021, the trends of the monthly SMR-S of males and females in 2021 were different from those in 2020, but similar to the trends of all 47 prefectures (Table 1, Figures 2 and 3). The monthly SMR-S of males and females during the second stay-home order increased, whereas the SMR-S of both males and females during the third stay-home order was almost equal to that in the pre-pandemic period (Table 1, Figures 2 and 3).

Therefore, fluctuations in suicide mortality in metropolitan regions were quite different in all 47 prefectures, since no decrease in the suicide mortality of males and females in metropolitan regions during the first stay-home order was observed. Over the past decade, monthly suicide mortality in Japan has been high in the first quarter and decreasing in the following periods [13,18,48]; however, the fluctuation pattern of suicide mortality in 2020 was a variation against the traditional pattern, as it increased in the third and fourth quarters (Figures 1 and 2).

Hierarchical linear model analysis detected a significant impact of SCR, SMR-C, and DCP on SMR-S during the pandemic period (between March 2020 and June 2021). The SMR-S of females in both all 47 prefectures and metropolitan regions was negatively related to SCR and SMR-C, but positively related to DCP; however, the SMR-S of males in both all 47 prefectures and metropolitan regions was not related to SCR, SMR-C, or DCP. These results suggest that the increase in patients with COVID-19 (SCR) and victims of COVID-19 (SMR-C) contributed to a reduction in the suicide mortality of females, but the prolongation of the pandemic led to increased female suicide mortality. The suicide mortality of males, on the other hand, is probably less sensitive to the influence of any data associated with COVID-19.

3.2. Suicide Mortality Disaggregated by Age

To identify the major factors of decreasing SMR-S during the first stay-home order and increasing SMR-S between the first and second stay-home orders, SMR-S disaggregated by age and gender factors was analysed using linear mixed-effects and hierarchical linear models.

In all 47 prefectures, in spite of decreasing monthly SMR-S in males during the first stay-home order, increasing monthly SMR-S in males over 50 between the first and second stay-home orders was not detected. The annual SMR-S of elderly (70s and 80s) females in 2020 decreased, whereas the annual SMR-S of females younger than 40 increased compared to the average annual SMR-S of the pre-pandemic period (Figures 4 and 5). In metropolitan regions, decreasing and increasing annual SMR-S in 2020 was not observed except for the SMR-S of males in their 50s and 10s, respectively (Figures 4 and 5).

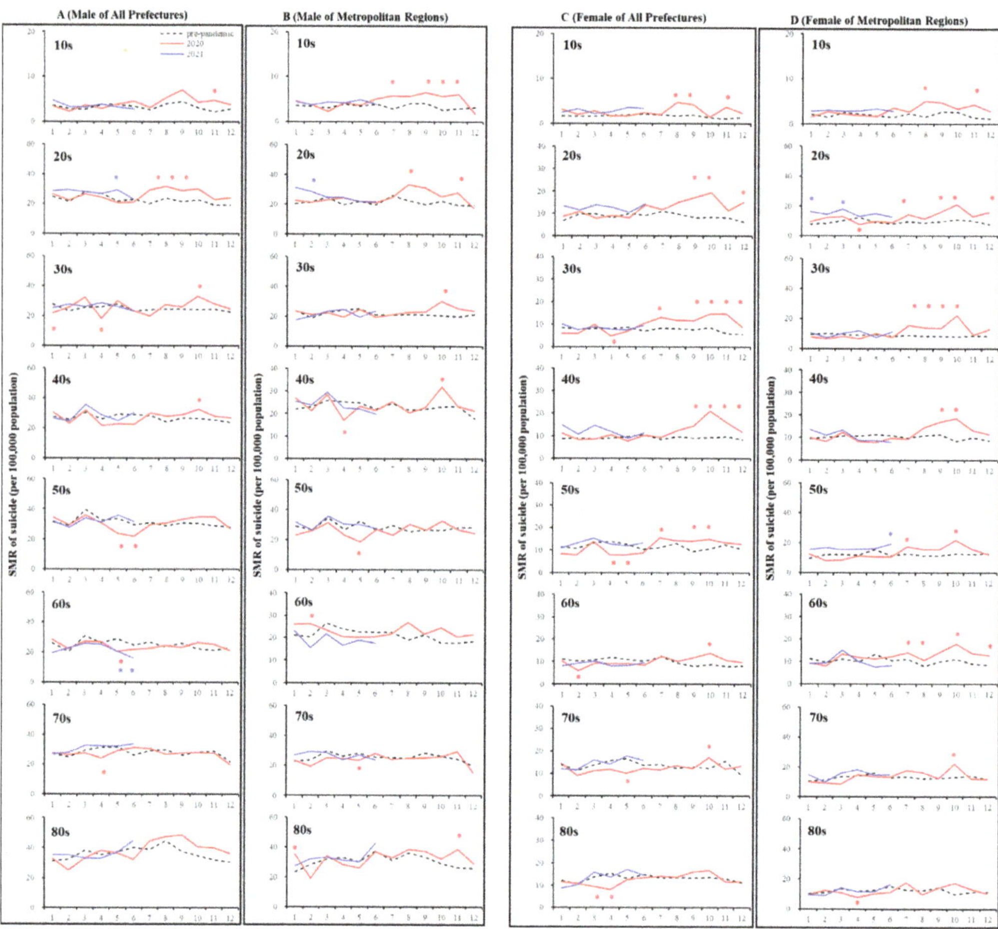

Figure 4. Temporal fluctuations in monthly SMR-S disaggregated by gender, age, and regional factors during the COVID-19 pandemic period (2020–2021) compared to the average of SMR-S during the same month during the pre-pandemic period (2017–2019). Dotted black, red, and blue lines indicate the average of SMR-S for (**A**) males from all 47 prefectures, (**B**) males from metropolitan regions, (**C**) females from all 47 prefectures, and (**D**) females from metropolitan regions. Ordinates indicate the SMR-S (per 100,000 people), and abscissas indicate the month. * $p < 0.05$, significant change using a linear mixed-effects model with Tukey's multiple comparison. Red and blue asterisks indicate significant changes in SMR-S in 2020 and 2021, respectively, compared to the average SMR-S of the same month during the pre-pandemic period.

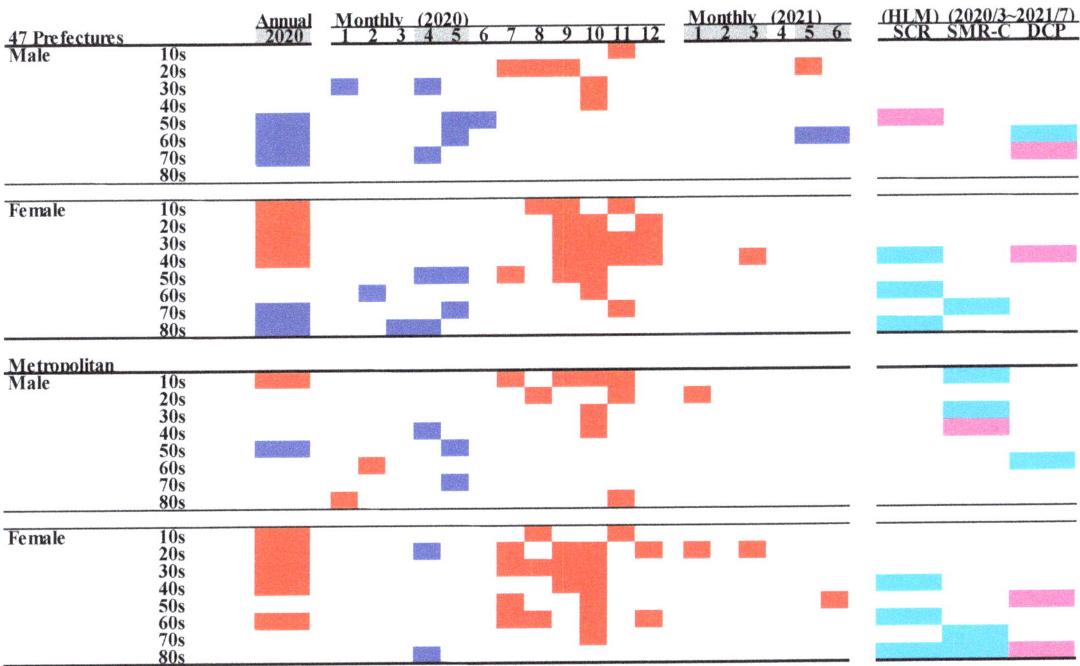

Figure 5. Comparison of annual and monthly SMR-S disaggregated by age during the pandemic period compared to the average SMR-S of the same month during the pre-pandemic period, and impact of COVID-19 pandemic on SMR-S in Japan. Blue and red columns indicate significant decreasing and increasing suicide mortality using a linear mixed-effects model with Tukey's multiple comparison ($p < 0.05$). Light blue and red columns indicate significant decreasing and increasing factors against SMR-S using hierarchical linear model with robust standard error ($p < 0.05$). SMR-S: standardised suicide mortality per 100,000 people. SCR: standardised infection with COVID-19 per 100,000 people. SMR-C: standardised mortality caused by COVID-19 per 100,000 people. DCP: during COVID-19 pandemic.

In all 47 prefectures, the kinetics of the monthly SMR-S of males and females decreased during the first stay-home order, but the monthly SMR-S of males and females was increased between the first and second stay-home orders compared to the average SMR-S for the same month during the pre-pandemic period (Figures 4 and 5). The tendency of decreasing SMR-S during the first stay-home order was more predominant in older populations in both males (30s–70s) and females (50s–80s) compared to younger populations. Increasing SMR-S between the first and second stay-home orders was detected in younger populations in males (10s–40s), but not in older males (50s–80s) (Figures 4 and 5). In contrast, a decrease during the first stay-home order and an increase between the first and second stay-home orders in the SMR-S of females were detected in a wide range of ages of females (30s–70s), but a decrease during the first stay-home order and an increase between the first and second stay-home orders in SMR-S were not observed in younger (10s and 20s) and elderly (80s) females, respectively (Figures 4 and 5).

In metropolitan regions, the age-dependent fluctuations of males' SMR-S were more pronounced, since the fluctuation of younger males (10s–30s) increased between the first and second stay-home orders without a decrease during the first stay-home order. Contrary to males, an increased SMR-S of wide-range-age females was observed, whereas a decrease in the SMR-S of females during the first stay-home order was not detected except for females in their 20s and 80s (Figures 4 and 5).

In SMR-S disaggregated by only the gender factor, the polarised responses of SMR-S to COVID-19-related data, SCR, SMR-C, and DCP were slightly weakened, using SMR-S disaggregated by gender with age factors by the hierarchical linear model (Figure 5). A positive relationship between SMR-S and DCP was still detected in females over 30 in all 47 prefectures, whereas a negative impact of SCR and SMR-C on SMR-S was not detected in females except for those in their 70s (Figure 5). Furthermore, in metropolitan regions, a positive impact of DCP on SMR-S was detected in only females in their 40s and 80s, whereas a negative impact of SCR and SMR-C on SMR-S was detected in females in their 60s–80s. In all 47 prefectures and metropolitan regions, a significant responsiveness of males' SMR-S to COVID-19-related data, SCR, SMR-C, and DCP was also detected, but with less consistent results (Figure 5). Nevertheless, the impact of COVID-19-related data, SCR, SMR-C, and DCP on the SMR-S of males was negligible compared to that of females.

3.3. Suicide Mortality Disaggregated by Household Condition

In all 47 prefectures, the annual SMR-S in 2020 for multiple-person household resident males and females decreased and increased, respectively; however, the SMR-S for one-person household resident males nor females did not change (Figures 6 and 7). The monthly SMR-S of multiple-person household resident males and females decreased during the first stay-home order (Figures 6 and 7). Contrary to multiple-person residents, the monthly SMR-S of one-person household residents (both males and females) did not change during the first stay-home order; however, between the first and second stay-home orders, the monthly SMR-S of multiple- and one-person resident males and females increased (Figures 6 and 7). The monthly SMR-S of one-person household resident males and females increased during the second or third stay-home order, whereas an increase was not detected in multiple-person household resident males or females during 2021 (Figures 6 and 7).

In metropolitan regions, a decrease in the monthly SMR-S of multiple- and one-person households resident males and females was not observed during the first stay-home order (Figure 5). Between the first and second stay-home orders, the monthly SMR-S of multiple- and one-person household resident males and females increased. Similar to all 47 prefectures, the monthly SMR-S of one-person household resident males and females increased during the second stay-home order. The monthly SMR-S of multiple-person household resident females increased in 2021, but that of males did not in 2021 (Figure 5).

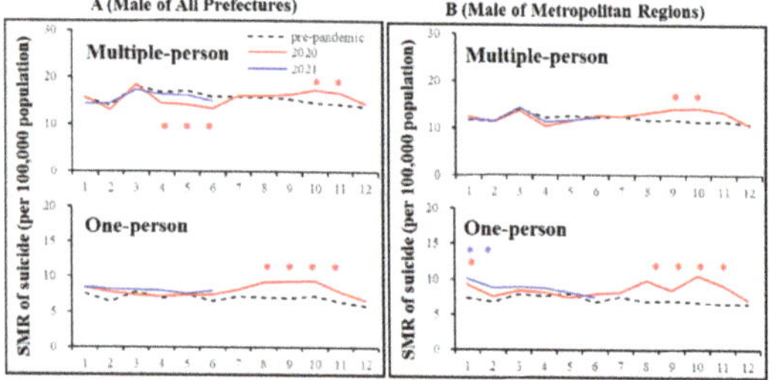
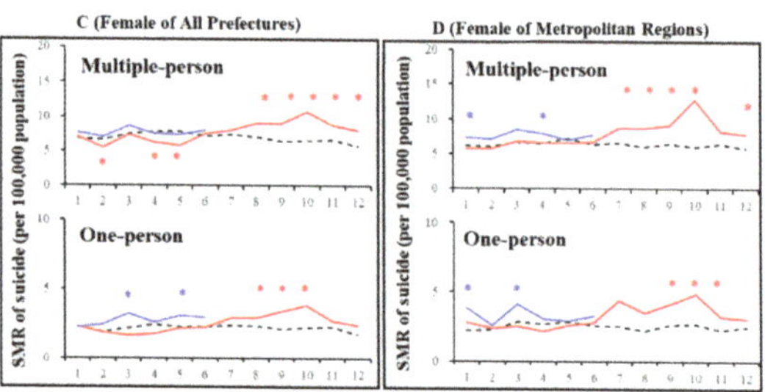

Figure 6. Temporal fluctuations in monthly SMR-S disaggregated by gender, household condition, and regional factors during the COVID-19 pandemic period (2020–2021) compared to the average of SMR-S for the same month during the pre-pandemic period (2017–2019). Dotted black, red, and blue lines indicate the average of SMR-S for (**A**) males from all 47 prefectures, (**B**) males from metropolitan regions, (**C**) females from all 47 prefectures and (**D**) females from metropolitan regions. Ordinates indicate the SMR-S (per 100,000 people), and abscissas indicate the month. * $p < 0.05$, significant change using a linear mixed-effects model with Tukey's multiple comparison. Red and blue asterisks indicate significant changes in SMR-S in 2020 and 2021, respectively, compared to the average SMR-S of the same month during the pre-pandemic period.

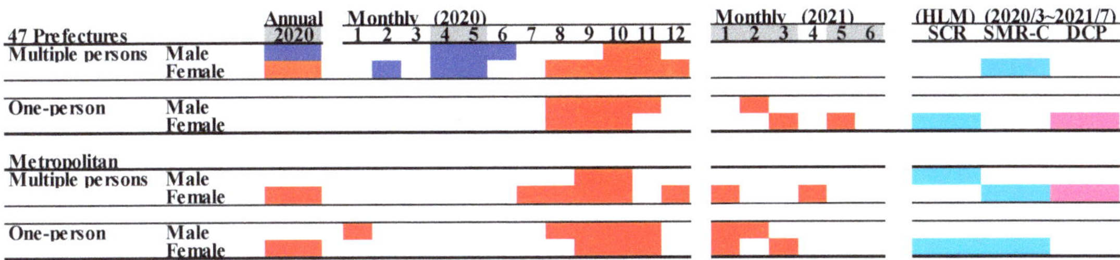

Figure 7. Comparison of annual and monthly SMR-S disaggregated by household condition during the pandemic period to average SMR-S of the same month during the pre-pandemic period, and impact of COVID-19 pandemic on SMR-S in Japan. Blue and red columns indicate significant decreasing and increasing suicide mortality using a linear mixed-effects model with Tukey's multiple comparison ($p < 0.05$). Light blue and red columns indicate significant decreasing and increasing factors against SMR-S using hierarchical linear model with robust standard error ($p < 0.05$). SMR-S: standardised suicide mortality per 100,000 people. SCR: standardised infection with COVID-19 per 100,000 people. SMR-C: standardised mortality caused by COVID-19 per 100,000 people. DCP: during COVID-19 pandemic.

The hierarchical linear model detected a significantly positive impact of DCP on the SMR-S of all females. There was a negative impact of SMR-C on the SMR-S of multiple-person household resident females in both all 47 prefectures and metropolitan regions, and one-person household resident females in metropolitan regions; however, SCR was negatively related to only the SMR-S of one-person household resident females in all 47 prefectures. The SMR-S of multiple-person household resident males in metropolitan regions was negatively related to SCR, whereas a relationship of other SMR-S of males with COVID-19-related data was not detected (Figure 5).

3.4. Suicide Mortality Disaggregated by Suicide Means

Out of five major suicide methods (hanging, poisoning, charcoal burning, jumping, and throwing), the SMR-S for hanging was specifically increased during the COVID-19 pandemic period compared to in the pre-pandemic period (Figures 8 and 9). The annual hanging SMR-S of females in all 47 prefectures and metropolitan regions specifically increased, but significant changes in other SMR-S were not detected (Figures 8 and 9). Furthermore, an increasing monthly SMR-S between the first and second stay-home orders was detected in the hanging suicide mortality of both males and females in all 47 prefectures and metropolitan regions; however, consistent changes in the other monthly SMR-S of poisoning, charcoal burning, jumping, and throwing were not observed (Figures 8 and 9). Increasing and decreasing SMR-S during and after the first stay-home order, respectively, of other suicide means (poisoning, charcoal burning, jumping, and throwing) were also detected, but these fluctuations were sporadic and nonpersistent. DCP and SMR-C were positively and negatively related to the hanging SMR-S of females in both all 47 prefectures and metropolitan regions. The hanging SMR-S of females in both all 47 prefectures and metropolitan regions displayed a persistent increase from the second half of 2020 to the first quarter of 2021 (Figures 8 and 9).

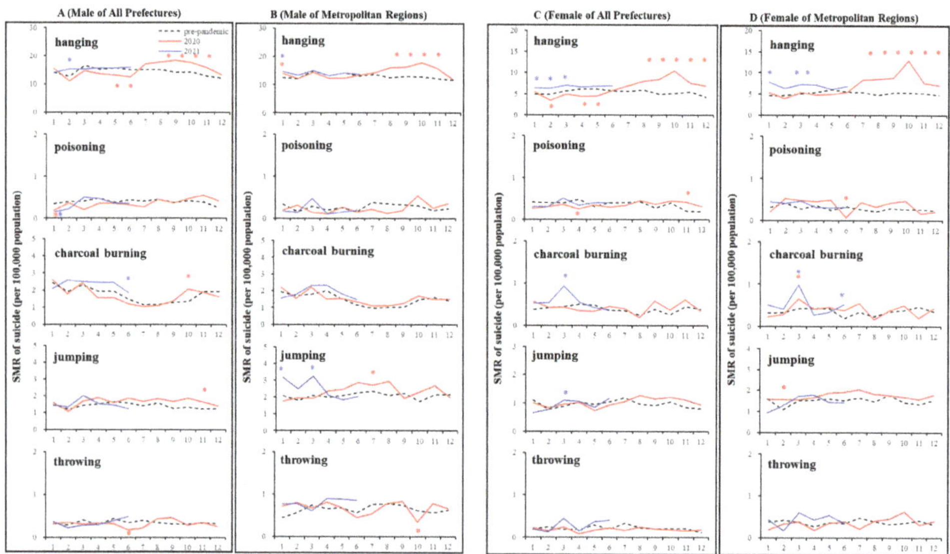

Figure 8. Temporal fluctuations in monthly SMR-S disaggregated by gender, suicidal means, and regional factors during COVID-19 pandemic period (2020–2021) compared to the average of SMR-S in the same month during the pre-pandemic period (2017–2019). Dotted black, red, and blue lines indicate the average of SMR-S for (**A**) males from all 47 prefectures, (**B**) males from metropolitan regions, (**C**) females from all 47 prefectures, and (**D**) females from metropolitan regions. Ordinates indicate the SMR-S (per 100,000 people), and abscissas indicate the month. * $p < 0.05$, significant change using a linear mixed-effects model with Tukey's multiple comparison. Red and blue asterisks indicate significant changes in SMR-S in 2020 and 2021, respectively, compared to the average SMR-S of the same month during the pre-pandemic period.

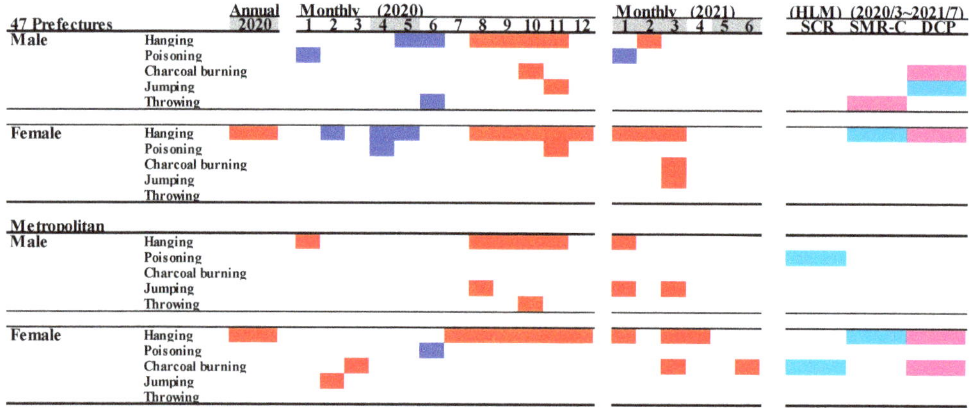

Figure 9. Comparison of annual and monthly SMR-S disaggregated by suicide means during the pandemic period to average SMR-S of the same month during pre-pandemic period, and impact of COVID-19 pandemic on SMR-S in Japan. Blue and red columns indicate significant decreasing and increasing suicide mortality using a linear mixed-effects model with Tukey's multiple comparison ($p < 0.05$). Light blue and red columns indicate significant decreasing and increasing factors against SMR-S using hierarchical linear model with robust standard error ($p < 0.05$). SMR-S: standardised suicide mortality per 100,000 people. SCR: standardised infection with COVID-19 per 100,000 people. SMR-C: standardised mortality caused by COVID-19 per 100,000 people. DCP: during COVID-19 pandemic.

3.5. Suicide Mortality by Motive

In Japan, the most dominant suicidal motive was health-related problems (for males in order: health > economy > family > employment > romance > school; for females in order: health > family > economy > employment > romance > school) [23,24,42]. The annual SMR-S of males caused by a health- and economy-related motive in 2020 decreased, but other SMR-S caused by family-, employment-, romance-, and school-related motives did not change. The annual SMR-S of females caused by employment- and school-related motives in 2020 increased, but other SMR-S caused by family-, health-, economy-, and school-related motives did not change. In metropolitan regions, the annual SMR-S of males disaggregated by motives did not change. The annual SMR-S of females caused by employment- and romance-related motives increased, but other SMR-S disaggregated by motives did not change in metropolitan regions.

In all 47 prefectures, the monthly SMR-S of males caused by health-, economy-, and employment-related motives decreased during the first stay-home order. The SMR-S of females caused by family- and economy-related motives also decreased, whereas that caused by health-related motives did not during the first stay-home order (Figures 10 and 11). Contrary to during the first stay-home order, between the first and second stay-home orders, female SMR-S caused by health- and family-related motives increased. Male SMR-S caused by health- and employment-related motives transiently increased, but that caused by family- and economy-related motives did not change between the first and second stay-home orders (Figures 10 and 11).

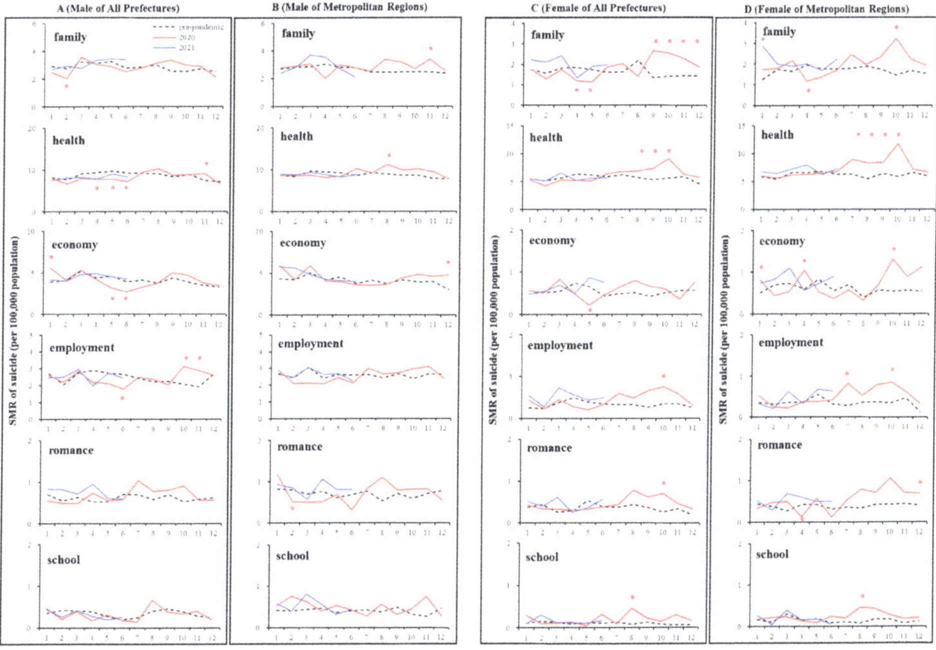

Figure 10. Temporal fluctuations in monthly SMR-S disaggregated by gender, suicidal reason, and regional factors during the COVID-19 pandemic period (2020–2021) compared to the average SMR-S of the same month during the pre-pandemic period (2017–2019). Dotted black, red, and blue lines indicate the average of SMR-S for (**A**) males from all 47 prefectures, (**B**) males from metropolitan regions, (**C**) females from all 47 prefectures, and (**D**) females from metropolitan regions. Ordinates indicate the SMR-S (per 100,000 people), and abscissas indicate the month. * $p < 0.05$, significant change using a linear mixed-effects model with Tukey's multiple comparison. Red and blue asterisks indicate significant changes in SMR-S in 2020 and 2021, respectively, compared to the average SMR-S of the same month during the pre-pandemic period.

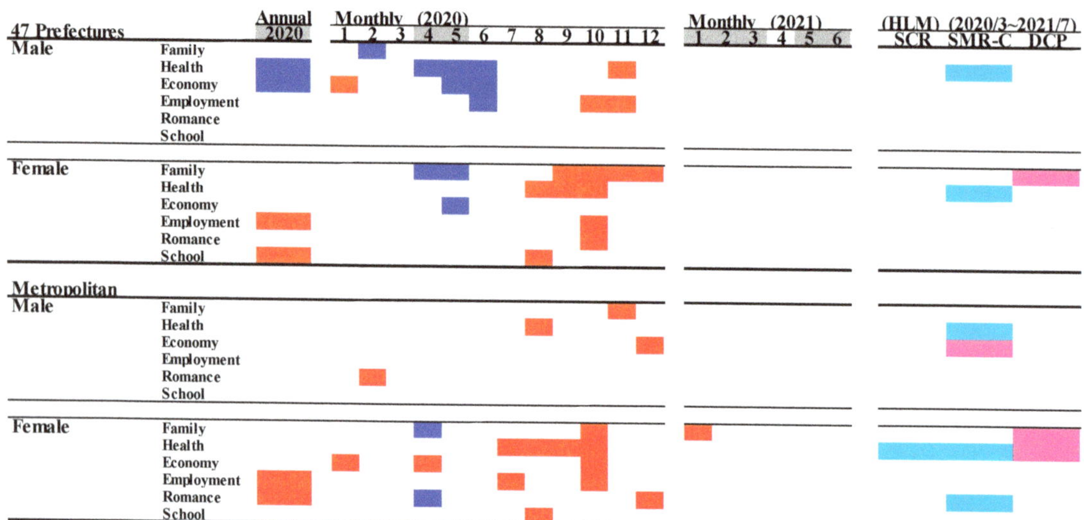

Figure 11. Comparison of annual and monthly SMR-S caused by suicidal motives during the pandemic period to average SMR-S of the same month during the pre-pandemic period, and impact of COVID-19 pandemic on SMR-S in Japan. Blue and red columns indicate significant decreasing and increasing suicide mortality using a linear mixed-effects model with Tukey's multiple comparison ($p < 0.05$). Light blue and red columns indicate significant decreasing and increasing factors against SMR-S using hierarchical linear model with robust standard error ($p < 0.05$). SMR-S: standardised suicide mortality per 100,000 people. SCR: standardised infection with COVID-19 per 100,000 people. SMR-C: standardised mortality caused by COVID-19 per 100,000 people. DCP: during COVID-19 pandemic.

In metropolitan regions, significant decreases in the monthly SMR-S of males caused by family-, health-, economy-, employment-, romance-, and school-related motives were not observed during the first stay-home order (Figures 10 and 11). In the dominant SMR-S of females, the monthly SMR-S caused by family- and romance-related motives also decreased, whereas that caused by economy-related motives was unexpectedly increased during the first stay-home order. Between the first and second stay-home orders, the monthly SMR-S of males caused by health-related motives was increased, and that caused by family-, economy-, and employment-related motives transiently increased. Male SMR-S caused by health-, family-, and economy-related motives transiently increased (Figures 10 and 11).

In all 47 prefectures, the hierarchical linear model detected a significant negative impact of SMR-C on the SMR-S of males and females caused by a health-related motive. The female SMR-S caused by a family-related motive was positively related to DCP (Figure 11). In metropolitan regions, the SMR-S of males caused by health-related motives was negatively related to SMR-C. Both SCR and SMR-C were negatively related to the SMR-S of females caused by health-related motives, and DCP was positively related to SMR-S caused by family- and health-related motives.

The SMR-S of females caused by economy-related motives in metropolitan regions was the sole increasing factor during the first stay-home order (Figures 10 and 11). Therefore, we reanalysed the SMR-S of males and females disaggregated by five major motives. In all 47 prefectures, the SMR-S of males and females caused by economy-related motives was decreased during the first stay-home order, and increased between the first and second stay-home orders (Figure 12). In contrast, the SMR-S of males and females caused by economy-related motives was increased in May–June 2020 and January–February 2021 without decreasing in the other months (Figure 12).

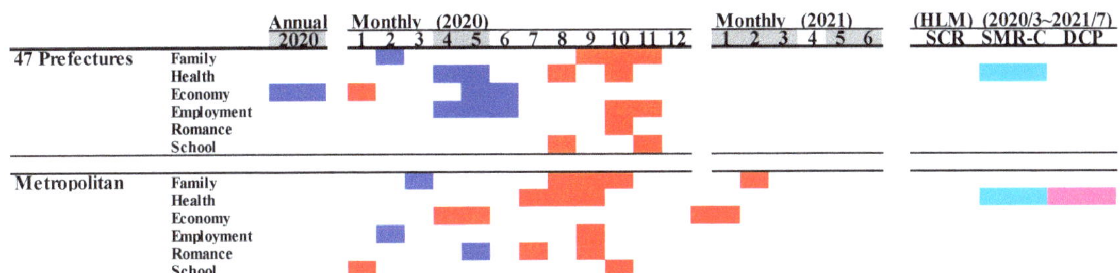

Figure 12. Comparison of annual and monthly SMR-S of males and females caused by suicidal motives during the pandemic period compared to average SMR-S of the same month during the pre-pandemic period, and impact of COVID-19 pandemic on SMR-S in Japan. Blue and red columns indicate significant decreasing and increasing suicide mortality using a linear mixed-effects model with Tukey's multiple comparison ($p < 0.05$). Light blue and red columns indicate significant decreasing and increasing factors against SMR-S using hierarchical linear model with robust standard error ($p < 0.05$). SMR-S: standardised suicide mortality per 100,000 people. SCR: standardised infection with COVID-19 per 100,000 people. SMR-C: standardised mortality caused by COVID-19 per 100,000 people. DCP: during COVID-19 pandemic.

4. Discussion

4.1. Overall Fluctuations of Suicide Mortality during COVID-19 Pandemic in Japan

The present study identified who died by suicide, when, why, and how during the COVID-19 pandemic period using a linear mixed-effects model and a hierarchical linear model with robust standard error. Overall fluctuations of suicide mortalities of males and females in 2020 led to a decrease during the first stay-home order (April–May 2020), and an increase between first and second stay-home orders (August–December 2020). However, the fluctuations in the suicide mortality of males and females appeared to be stabilised in the first half of 2021, including during the second (January–March 2021) and third (April–Jun 2021) stay-home orders. Contrary to national trends of suicide mortality, there was a lack of a decrease in the suicide mortality of males and females during the first stay-home order in five metropolitan regions; however, fluctuations in the suicide mortality of males and females in metropolitan regions also appeared to be stabilising in the first half of 2021. Therefore, fluctuations in suicide mortality in 2020, decreasing during the first stay-home order and increasing between the first and second stay-home orders, probably constituted a specific pattern. A recent study revealed that the risk of suicide in 21 countries could not be detected during the early COVID-19 pandemic periods. Initially, the first stay-home order was announced to metropolitan regions due to the spread COVID-19 in these regions in April 2020, but a short-term stay-home order was then announced to all 47 prefectures in May 2020 to prevent the spread of COVID-19 to rural regions induced by long holidays in May (Golden Week). Interpreted on the basis of previous natural disaster cases, the reduction in SMR-S in all 47 prefectures during the first stay-home order might be similar to the "honeymoon period" phenomenon [37,38]. However, the smaller reduction in SMR-S in the metropolitan regions during the first stay-home order cannot deny the possibility that the impact associated with COVID-19 spread or the first stay-home order on individuals in metropolitan regions was greater than that in other regions. The stabilisation of suicide mortality observed with each stay-home order may also suggest becoming accustomed to the pandemic; however, hierarchical linear model analysis detected an interaction between the negative effects of DCP (duration of the pandemic) and the positive effects of SMR-C (mortality caused by COVID-19) and SCR (infected population with COVID-19) on the suicide mortality of females, possibly neutralising each other. Therefore, suicide mortality was probably composed of various factors' interactions.

4.2. When and Who Died by Suicide

This characteristic could be detected even in the issue of who died by suicide and when. Regarding the analysis of all 47 prefectures, the suicide mortality of over 40s males decreased during the first stay-home order, and did not increase between the first and second stay-home orders, whereas the suicide mortality of young (10s–20s) males did not decrease during the first stay-home order, but increased between the first and second stay-home orders. Additionally, regional characteristics of the fluctuation in males' SMR-S in 2020 could not be detected. In the analysis of females from all 47 prefectures, the suicide mortality of young females (10s–20s) did not decrease during the first stay-home order, but increased between the first and second stay-home orders, similar to the SMR-S of young males. Contrary to males, the suicide mortality of females over 30 also decreased during the first stay-home order, but increased between the first and second stay-home orders, contrary to the suicide mortality of males over the age of 40. Regarding metropolitan regions, the SMR-S of males in metropolitan regions had a similar fluctuation pattern in all 47 prefectures; however, the SMR-S of females during the first stay-home order was minor, but increased between the first and second stay-home orders. Therefore, the combination of the increasing suicide mortality of females between the first and second stay-home orders and the less decreasing suicide mortality of females in metropolitan regions contributed to the increase in suicide mortality in 2020.

The gender-specific characteristics of SMR-S fluctuation between multiple-person household resident males and females could not be observed in all 47 prefectures, since the SMR-S of both males and females decreased during the first stay-home order and increased between the first and second stay-home orders. Contrarily, SMR-S fluctuations among one-person household resident males and females displayed a different pattern to those of multiple-person household residents, since no decrease in the suicide mortality of one-person household residents during the first stay-home order was found across all 47 prefectures. This lack of a decrease during the first stay-home order was also observed in both multiple-person and one-person household resident males and females in metropolitan regions. Therefore, the lack of a decrease in suicide mortalities among one-person household residents during the first stay-home order contributed to an increase in the annual suicide mortality of females in 2020.

4.3. When and Who Died by Suicide, and How

This characteristic could also be detected even in the issue of who died by suicide, when, and how. A consistent increase in hanging suicide could be detected between the first and second stay-home orders, irrespective of gender or region. Mass media frequently reported on celebrities dying by hanging suicide in July and September 2020. MHLW speculated that the increasing suicide mortality of females between the first and second stay-home orders was probably induced by these frequent reports of mass media [27,28]. In particular, these frequent reports from mass media deviated from the suicide reporting WHO guidelines "Preventing suicide: a resource for media professionals" [26]. However, the most dominant locations and tools of hanging suicide, which was the most common means of suicide in Japan [25,48], were people's homes and every-day items, such as belts, electric flex, rafters or beams, bannisters, hooks, doorknobs, and trees [25,49]. Initially, we considered that the increase in the length of staying at home due to the stay-home order was a dominant risk factor of hanging suicide; however, the period of increasing hanging suicide mortality was not during, but following the end of the stay-home order. Although it is impossible to detect the more detailed factors behind the increasing hanging suicides in this study, it is speculated that individuals who suspended hanging suicide by the first stay-home order did die by hanging suicide due to the end of the first stay-home order, since the prolongation of the pandemic was a risk for increasing the hanging suicide mortality of females. Further analysis to identify the background factors of increasing hanging suicide using various independent variables will be published to provide important findings.

4.4. Who Died by Suicide, When, and Why

This characteristic could also even be detected in the issue of who died by suicide, when, and why. The most predominant suicide mortality of both males and females was caused by health-related motives [23,24,42] and was suppressed by the increasing number of COVID-19 victims (SMR-C) detected by hierarchical linear model analysis. It is easy to interpret that the lack of increasing suicide mortality of males caused by health-related motives between the first and second stay-home orders (predominant in metropolitan regions as severe infected areas with COVID-19) was sufficient to offset other increased suicides of males. Although the suicide mortality of males caused by economy-related motives in all 47 prefectures decreased during the first stay-home order, the suicide mortality of females caused by economy-related motives in metropolitan regions was unexpected to be increased in April 2020. Furthermore, in spite of lacking a significant change in males in the metropolitan region, the suicide mortality of males and females caused by economy-related motives in metropolitan regions increased during May–June 2020. The postponement of the Tokyo 2020 Olympics due to the COVID-19 pandemic was decided on 24 March 2021. The postponement decision of the Tokyo 2020 Olympics did not publish the detailed postponement date with the possible cancellation [50]. The economic effect of the Tokyo 2020 Olympics was estimated to be at least 0.2–0.3% of GDP per year [51]. This effect is particularly concentrated in the capital metropolitan area, and it can be easily estimated that the economic damages due to postponement were also large in metropolitan areas. Therefore, Japanese socioeconomic deterioration status suffered due to both the stay-home order for the suppression of COVID-19 pandemic and the postponement of the Olympics in 2020. In other words, decreasing suicide mortality in metropolitan regions during the first stay-home order was slight compared to that in all 47 prefectures, and probably generated by characteristics of urban areas and the seriousness of the COVID-19 infection situation, and by being offset through economic damages due to the postponement of the Olympics. Detailed analysis shows that the interaction between the COVID-19 pandemic and the postponement of the Olympics on economic activity in 2020 plays important roles in the clarification of suicide mortality caused by economy-related motives in capital metropolitan areas.

4.5. Candidate Mechanisms of Specific Fluctuations of Suicide Mortality of Younger Populations

The present study indicated several possible factors of increasing female suicide mortality in 2020. We considered that the lack of decreasing suicide mortality of young populations and one-person household resident females during the first stay-home order, and the increase after the end of the first stay-home order were characteristic fluctuation patterns of suicide mortality in 2020. A report speculated that the reason for the increased suicide mortality of young females in Japan was that the stay-home order led to an increasing unemployment rate with a decrease in temporary employment via economic recession [18]. However, the rate and number of temporary jobs between the first and second stay-home orders were deteriorated compared to the pre-pandemic period (in 2019), but this tendency was not specific to young populations [52]. Increasing domestic violence against females was a possible reason for the increasing Japanese female suicide mortality in 2020 [18], whereas instances of domestic violence events in 2020 were fewer than those in 2019 [53]. Therefore, economic recession and increasing domestic violence probably did not play important roles in the fluctuations of suicide mortality of young populations or females in 2020.

Online communication has been a standard tool in the social landscape of young populations in recent years [54]. During the pandemic, online communication became indispensable. Online communication tools played important roles in preventing isolation and maintaining schooling opportunities during the stay-home order period. However, the lengthening duration of passive social media communication in young females was related to increasing depressive symptoms [55]. Recent studies suggested that young females felt less life satisfaction and increased conflict with their parents during the COVID-19 pandemic compared to young males [56,57]. Increased family contact during the stay-home order possibly relieved the stress of young females, including academic problems, resulting in mitigating the potential negative impacts of the COVID-19 pandemic [58,59]. Therefore, young one-person-household females have probably been suffering from a vicious cycle of increasing reliance on online communication without family support. The Japan Suicide Countermeasures Promotion Center reported the possibility that the increasing suicide mortality after the end of the first stay-home order could not be explained only by the frequent suicide reports from the mass media, but also due to the combination between the large spread of the words "suicide" in SNS and the increasing suicide reports from the mass media [60]. Therefore, it is undeniable that under increasing exposure to passive online communication, the spread of the word "suicide" in SNS probably increased the risk of suicide for young one-person household resident females.

4.6. Limitations

There were several limitations in this study. First, BDSR published the numbers of suicide victims disaggregated by occupation; however, SBMIAC did not publish the exact occupational population as a denominator. Second, BDSR did not also publish the annual and monthly suicide mortality disaggregated by motive and age or means and age. The present study could not identify detailed background factors of increasing female suicide mortality between the first and second stay-home orders induced by these two limitations. Third, the COVID-19 pandemic is not yet over, but its impact on suicide mortality using various economic, financial, medical, and welfare indicators and comparisons of longer-term surveys of suicide mortality between Japan and other countries could identify the detailed background associated with the COVID-19 pandemic.

5. Conclusions

Characteristics of fluctuations in suicide mortality in Japan during the COVID-19 pandemic were outlined. Suicide mortality decreased during the first stay-home order and increased after the end of the first stay-home order. Furthermore, the direct health hazard of COVID-19 itself functioned as a suicide suppressor, but the prolongation of the COVID-19 pandemic period contributed to the increasing suicide mortality of females. Contrary to nationwide fluctuation patterns of suicide mortality, the suicide mortality of both males and females in metropolitan regions did not have a decreasing phase during the first stay-home order. Other factors, females, adolescents, one-person household residents, and metropolitan areas were possible risks of increasing suicide mortality in 2020. Additionally, the postponement of the Tokyo 2020 Olympics attenuated the decreasing suicide mortality during the first stay-home order in metropolitan regions. Taken together with previous findings associated with socioeconomic and sociopsychological deterioration, in the context of Japan, a number of reports concerned the overheated reports of mass media, financial stress, and unemployment under the governmental COVID-19 pandemic restrictions. Although suicides in Japan might have had various influences associated with the COVID-19 pandemic, these influences were consistent with complicated reasons among direct and indirect factors associated with the pandemic. Although online communication tools are important for maintaining education opportunities and preventing isolation in young populations, online communication itself possibly promotes suicide in young populations. Therefore, the enhancement of online communication tools needs to be considered as a double-edged sword for young populations.

Author Contributions: Conceptualisation, M.O.; methodology, R.M.; validation, E.M. and K.F.; formal analysis, R.M. and M.O.; investigation, R.M. and M.O.; data curation, M.O.; writing—original draft preparation, T.S. and M.O.; writing—review and editing, M.O. All authors have read and agreed to the published version of the manuscript.

Funding: This study was supported by the Regional Suicide Countermeasures Emergency Enhancement Fund of Mie Prefecture (2021-40).

Institutional Review Board Statement: No patients or members of the public were involved in this study. Only secondary data were used for the analyses. This study was exempt from ethical approval by Mie University.

Informed Consent Statement: Not applicable.

Data Availability Statement: All data relevant to the study are included in the article. All raw data are available to any persons via Japanese national databases from the Statistics Bureau of the Ministry of Internal Affairs and Communications (SBMIAC), the Cabinet Office (CAO), and the Ministry of Health, Labour and Welfare (MHLW).

Conflicts of Interest: The authors declare no conflict of interest.

References

1. World Health Organization. Who Coronavirus Disease (COVID-19) Dashboard. Available online: https://covid19.who.int/ (accessed on 25 October 2021).
2. National Institute of Infectious Diseases. Report Week Correspondence Table. Available online: https://www.niid.go.jp/niid/ja/calendar.html (accessed on 1 September 2021).
3. Banerjee, D.; Kosagisharaf, J.R.; Sathyanarayana Rao, T.S. 'The dual pandemic' of suicide and COVID-19: A biopsychosocial narrative of risks and prevention. *Psychiatry Res.* **2021**, *295*, 113577. [CrossRef] [PubMed]
4. Gunnell, D.; Appleby, L.; Arensman, E.; Hawton, K.; John, A.; Kapur, N.; Khan, M.; O'Connor, R.C.; Pirkis, J.; Caine, E.D.J.T.L.P. Suicide risk and prevention during the COVID-19 pandemic. *Lancet Psychiatry* **2020**, *7*, 468–471. [CrossRef]
5. Kawohl, W.; Nordt, C.J.T.L.P. COVID-19, unemployment, and suicide. *Lancet Psychiatry* **2020**, *7*, 389–390. [CrossRef]
6. Reger, M.A.; Stanley, I.H.; Joiner, T.E. Suicide mortality and coronavirus disease 2019—a perfect storm? *JAMA Psychiatry* **2020**, *77*, 1093–1094. [CrossRef]
7. Wasserman, D.; Iosue, M.; Wuestefeld, A.; Carli, V. Adaptation of evidence-based suicide prevention strategies during and after the COVID-19 pandemic. *World Psychiatry* **2020**, *19*, 294–306. [CrossRef]
8. Chang, S.S.; Gunnell, D.; Sterne, J.A.; Lu, T.H.; Cheng, A.T. Was the economic crisis 1997–1998 responsible for rising suicide rates in East/Southeast Asia? A time-trend analysis for Japan, Hong Kong, South Korea, Taiwan, Singapore and Thailand. *Soc. Sci. Med.* **2009**, *68*, 1322–1331. [CrossRef]
9. Anagnostopoulos, D.C.; Giannakopoulos, G.; Christodoulou, N.G. The synergy of the refugee crisis and the financial crisis in Greece: Impact on mental health. *Int. J. Soc. Psychiatry* **2017**, *63*, 352–358. [CrossRef] [PubMed]
10. Pirkis, J.; John, A.; Shin, S.; DelPozo-Banos, M.; Arya, V.; Analuisa-Aguilar, P.; Appleby, L.; Arensman, E.; Bantjes, J.; Baran, A.; et al. Suicide trends in the early months of the COVID-19 pandemic: An interrupted time-series analysis of preliminary data from 21 countries. *Lancet Psychiatry* **2021**, *8*, 579–588. [CrossRef]
11. John, A.; Eyles, E.; Webb, R.T.; Okolie, C.; Schmidt, L.; Arensman, E.; Hawton, K.; O'Connor, R.C.; Kapur, N.; Moran, P.; et al. The impact of the COVID-19 pandemic on self-harm and suicidal behaviour: Update of living systematic review. *F1000Research* **2020**, *9*, 1097. [CrossRef] [PubMed]
12. Rogers, J.P.; Chesney, E.; Oliver, D.; Begum, N.; Saini, A.; Wang, S.; McGuire, P.; Fusar-Poli, P.; Lewis, G.; David, A.S. Suicide, self-harm and thoughts of suicide or self-harm in infectious disease epidemics: A systematic review and meta-analysis. *Epidemiol. Psychiatr. Sci.* **2021**, *30*, e32. [CrossRef]
13. Ministry of Health, Law. Basic Data on Suicide in the Region. Available online: https://www.mhlw.go.jp/stf/seisakunitsuite/bunya/0000140901.html (accessed on 1 September 2021).
14. Fushimi, M. The importance of studying the increase in suicides and gender differences during the COVID-19 pandemic. *QJM Int. J. Med.* **2021**. [CrossRef]
15. Sakamoto, H.; Ishikane, M.; Ghaznavi, C.; Ueda, P. Assessment of Suicide in Japan during the COVID-19 Pandemic vs Previous Years. *JAMA Netw. Open* **2021**, *4*, e2037378. [CrossRef]
16. Nomura, S.; Kawashima, T.; Harada, N.; Yoneoka, D.; Tanoue, Y.; Eguchi, A.; Gilmour, S.; Kawamura, Y.; Hashizume, M. Trends in suicide in Japan by gender during the COVID-19 pandemic, through December 2020. *Psychiatry Res.* **2021**, *300*, 113913. [CrossRef] [PubMed]
17. Seposo, X.T. COVID-19 threatens decade-long suicide initiatives in Japan. *Asian J. Psychiatr.* **2021**, *60*, 102660. [CrossRef]
18. Eguchi, A.; Nomura, S.; Gilmour, S.; Harada, N.; Sakamoto, H.; Ueda, P.; Yoneoka, D.; Tanoue, Y.; Kawashima, T.; Hayashi, T.I.; et al. Suicide by gender and 10-year age groups during the COVID-19 pandemic vs previous five years in Japan: An analysis of national vital statistics. *Psychiatry Res.* **2021**, *305*, 114173. [CrossRef] [PubMed]
19. Tanaka, T.; Okamoto, S. Increase in suicide following an initial decline during the COVID-19 pandemic in Japan. *Nat. Hum. Behav.* **2021**, *5*, 229–238. [CrossRef] [PubMed]
20. Kino, S.; Jang, S.N.; Gero, K.; Kato, S.; Kawachi, I. Age, period, cohort trends of suicide in Japan and Korea (1986–2015): A tale of two countries. *Soc. Sci. Med.* **2019**, *235*, 112385. [CrossRef]
21. Kato, R.; Okada, M. Can Financial Support Reduce Suicide Mortality Rates? *Int. J. Environ. Res. Public Health* **2019**, *16*, 4797. [CrossRef] [PubMed]

22. Okada, M.; Hasegawa, T.; Kato, R.; Shiroyama, T. Analysing regional unemployment rates, GDP per capita and financial support for regional suicide prevention programme on suicide mortality in Japan using governmental statistical data. *BMJ Open* **2020**, *10*, e037537. [CrossRef]

23. Nakamoto, M.; Nakagawa, T.; Murata, M.; Okada, M. Impacts of Dual-Income Household Rate on Suicide Mortalities in Japan. *Int. J. Environ. Res. Public Health* **2021**, *18*, 5670. [CrossRef]

24. Shiroyama, T.; Fukuyama, K.; Okada, M. Effects of Financial Expenditure of Prefectures/Municipalities on Regional Suicide Mortality in Japan. *Int. J. Environ. Res. Public Health* **2021**, *18*, 8639. [CrossRef]

25. Hasegawa, T.; Matsumoto, R.; Yamamoto, Y.; Okada, M. Analysing effects of financial support for regional suicide prevention programmes on methods of suicide completion in Japan between 2009 and 2018 using governmental statistical data. *BMJ Open* **2021**, *11*, e049538. [PubMed]

26. Ministry of Health, Labor and Welfare. Preventing Suicide: A Resource for Media Professionals-Update 2017. Available online: https://www.who.int/mental_health/suicide-prevention/resource_booklet_2017/en/ (accessed on 1 December 2019).

27. Ministry of Health, Labor and Welfare. Survey Study on the Actual Conditions of Young Carers. Available online: https://elaws.e-gov.go.jp/search/elawsSearch/elaws_search/lsg0500/detail?lawId=501AC1000000032_20190912_000000000 000000&openerCode=1 (accessed on 1 May 2021).

28. Ministry of Health, Labor and Welfare. Calling Attention When Reporting on Celebrity Suicide. Available online: https://www.mhlw.go.jp/stf/seisakunitsuite/bunya/hukushi_kaigo/seikatsuhogo/jisatsu/who_tebiki.html (accessed on 25 October 2021).

29. Wasserman, I.M. The impact of epidemic, war, prohibition and media on suicide: United States, 1910–1920. *Suicide Life Threat. Behav.* **1992**, *22*, 240–254.

30. Liang, S.T.; Liang, L.T.; Rosen, J.M. COVID-19: A comparison to the 1918 influenza and how we can defeat it. *Postgrad. Med. J.* **2021**, *97*, 273–274. [CrossRef]

31. Chan, S.M.; Chiu, F.K.; Lam, C.W.; Leung, P.Y.; Conwell, Y. Elderly suicide and the 2003 SARS epidemic in Hong Kong. *Int. J. Geriatr. Psychiatry* **2006**, *21*, 113–118. [CrossRef] [PubMed]

32. Cheung, Y.T.; Chau, P.H.; Yip, P.S. A revisit on older adults suicides and Severe Acute Respiratory Syndrome (SARS) epidemic in Hong Kong. *Int. J. Geriatr. Psychiatry* **2008**, *23*, 1231–1238. [CrossRef] [PubMed]

33. Yip, P.S.; Cheung, Y.T.; Chau, P.H.; Law, Y.W. The impact of epidemic outbreak: The case of severe acute respiratory syndrome (SARS) and suicide among older adults in Hong Kong. *Crisis* **2010**, *31*, 86–92. [CrossRef]

34. Zortea, T.C.; Brenna, C.T.A.; Joyce, M.; McClelland, H.; Tippett, M.; Tran, M.M.; Arensman, E.; Corcoran, P.; Hatcher, S.; Heisel, M.J.; et al. The Impact of Infectious Disease-Related Public Health Emergencies on Suicide, Suicidal Behavior, and Suicidal Thoughts. *Crisis* **2020**, 1–14. [CrossRef]

35. Fire and Disaster Management Agency. 2014 Firefighting White Paper. Available online: https://www.fdma.go.jp/publication/#whitepaper (accessed on 1 June 2021).

36. Matsubayashi, T.; Sawada, Y.; Ueda, M. Natural disasters and suicide: Evidence from Japan. *Soc. Sci. Med.* **2013**, *82*, 126–133. [CrossRef]

37. Madianos, M.G.; Evi, K. Trauma and natural disaster: The case of earthquakes in Greece. *J. Loss Trauma* **2010**, *15*, 138–150. [CrossRef]

38. Gordon, K.H.; Bresin, K.; Dombeck, J.; Routledge, C.; Wonderlich, J.A.J.C. The impact of the 2009 Red River Flood on interpersonal risk factors for suicide. *Crisis* **2011**, *32*, 52–55. [CrossRef]

39. Farooq, S.; Tunmore, J.; Ali, W.; Ayub, M. Suicide, self-harm and suicidal ideation during COVID-19: A systematic review. *Psychiatry Res.* **2021**, *306*, 114228. [CrossRef]

40. John, A.; Pirkis, J.; Gunnell, D.; Appleby, L.; Morrissey, J. Trends in suicide during the COVID-19 pandemic. *BMJ* **2020**, *371*, m4352. [CrossRef]

41. Statistics Bureau of the Ministry of Internal Affairs and Communications. Surveys of Population, Population Change and the Number of Households Based on the Basic Resident Registration. Available online: https://www.e-stat.go.jp/stat-search/files?page=1&toukei=00200241&tstat=000001039591 (accessed on 1 September 2021).

42. Nakano, T.; Hasegawa, T.; Okada, M. Analysing the Impacts of Financial Support for Regional Suicide Prevention Programmes on Suicide Mortality Caused by Major Suicide Motives in Japan Using Statistical Government Data. *Int. J. Environ. Res. Public Health* **2021**, *18*, 3414. [CrossRef]

43. Shiratori, Y.; Tachikawa, H.; Nemoto, K.; Endo, G.; Aiba, M.; Matsui, Y.; Asada, T. Network analysis for motives in suicide cases: A cross-sectional study. *Psychiatry Clin. Neurosci.* **2014**, *68*, 299–307. [CrossRef] [PubMed]

44. Idogawa, M.; Tange, S.; Nakase, H.; Tokino, T. Interactive Web-based Graphs of Coronavirus Disease 2019 Cases and Deaths per Population by Country. *Clin. Infect. Dis.* **2020**, *71*, 902–903. [CrossRef]

45. Fukuyama, K.; Kato, R.; Murata, M.; Shiroyama, T.; Okada, M. Clozapine Normalizes a Glutamatergic Transmission Abnormality Induced by an Impaired NMDA Receptor in the Thalamocortical Pathway via the Activation of a Group III Metabotropic Glutamate Receptor. *Biomolecules* **2019**, *9*, 234. [CrossRef] [PubMed]

46. Nakano, T.; Hasegawa, T.; Suzuki, D.; Motomura, E.; Okada, M. Amantadine Combines Astroglial System Xc(-) Activation with Glutamate/NMDA Receptor Inhibition. *Biomolecules* **2019**, *9*, 191. [CrossRef] [PubMed]

47. Okada, M.; Fukuyama, K.; Kawano, Y.; Shiroyama, T.; Ueda, Y. Memantine protects thalamocortical hyper-glutamatergic transmission induced by NMDA receptor antagonism via activation of system xc. *Pharmacol. Res. Perspect.* **2019**, *7*, e00457. [CrossRef]
48. Ministry of Health, Labor and Welfare. 2020 White Paper on Suicide Prevention. Available online: https://www.mhlw.go.jp/stf/seisakunitsuite/bunya/hukushi_kaigo/seikatsuhogo/jisatsu/jisatsuhakusyo2020.html (accessed on 1 June 2021).
49. Gunnell, D.; Bennewith, O.; Hawton, K.; Simkin, S.; Kapur, N. The epidemiology and prevention of suicide by hanging: A systematic review. *Int. J. Epidemiol.* **2005**, *34*, 433–442. [CrossRef]
50. IOC. Joint Statement from the International Olympic Committee and the Tokyo 2020 Organising Committee. Available online: https://olympics.com/ioc/news/joint-statement-from-the-international-olympic-committee-and-the-tokyo-2020-organising-committee (accessed on 1 September 2021).
51. Nagata, M.; Ojima, M.; Kurachi, T.; Miura, H.; Kawamoto, T. Economic effects of the 2020 Tokyo Olympics (Japanese). *Bank Jpn. Rep. Res. Pap.* **2015**, *12*, 1–11.
52. Statistics Bureau of the Ministry of Internal Affairs and Communications. Labor Force Survey. Available online: https://www.stat.go.jp/data/roudou/pref/index.html (accessed on 26 October 2021).
53. Agency, N.P. Criminal Statistics: Stoker and Domestic Violence. Available online: https://www.npa.go.jp/publications/statistics/safetylife/dv.html (accessed on 26 October 2021).
54. Odgers, C.L.; Schueller, S.M.; Ito, M. Screen time, social media use, and adolescent development. *Annu. Rev. Dev. Psychol.* **2020**, *2*, 485–502. [CrossRef]
55. Ellis, W.E.; Dumas, T.M.; Forbes, L.M. Physically isolated but socially connected: Psychological adjustment and stress among adolescents during the initial COVID-19 crisis. *Can. J. Behav. Sci. /Rev. Can. Des. Sci. Du Comport.* **2020**, *52*, 177. [CrossRef]
56. Kapetanovic, S.; Gurdal, S.; Ander, B.; Sorbring, E. Reported changes in adolescent psychosocial functioning during the COVID-19 outbreak. *Adolescents* **2021**, *1*, 10–20. [CrossRef]
57. Magson, N.R.; Freeman, J.Y.A.; Rapee, R.M.; Richardson, C.E.; Oar, E.L.; Fardouly, J. Risk and Protective Factors for Prospective Changes in Adolescent Mental Health during the COVID-19 Pandemic. *J. Youth Adolesc.* **2021**, *50*, 44–57. [CrossRef] [PubMed]
58. Fegert, J.M.; Vitiello, B.; Plener, P.L.; Clemens, V. Challenges and burden of the Coronavirus 2019 (COVID-19) pandemic for child and adolescent mental health: A narrative review to highlight clinical and research needs in the acute phase and the long return to normality. *Child. Adolesc. Psychiatry Ment. Health* **2020**, *14*, 20. [CrossRef] [PubMed]
59. Hoekstra, P.J. Suicidality in children and adolescents: Lessons to be learned from the COVID-19 crisis. *Eur. Child. Adolesc. Psychiatry* **2020**, *29*, 737–738. [CrossRef]
60. Japan Suicide Countermeasures Promotion Center. Enlightenment/Recommendation. Available online: https://jscp.or.jp/action/jisatsu_benkyokai_report0810.html (accessed on 26 October 2021).

Journal of
Clinical Medicine

Review

Incidence of Mortality, Acute Kidney Injury and Graft Loss in Adult Kidney Transplant Recipients with Coronavirus Disease 2019: Systematic Review and Meta-Analysis

Jia-Jin Chen [1,†], George Kuo [1,†], Tao Han Lee [1], Huang-Yu Yang [1,2], Hsin Hsu Wu [1,2], Kun-Hua Tu [1,2] and Ya-Chung Tian [1,2,*]

1   Department of Nephrology, Chang Gung Memorial Hospital, Linkou Main Branch, Taoyuan City 33305, Taiwan; raymond110234@hotmail.com (J.-J.C.); b92401107@gmail.com (G.K.); kate0327@hotmail.com (T.H.L.); hyyang01@gmail.com (H.-Y.Y.); tomwu38@gmail.com (H.H.W.); christopher.tu@gmail.com (K.-H.T.)
2   Department of Nephrology, Kidney Research Center, Chang Gung Memorial Hospital, Taoyuan City 33305, Taiwan
*   Correspondence: dryctian@cgmh.org.tw
†   Co-first authors.

Citation: Chen, J.-J.; Kuo, G.; Lee, T.H.; Yang, H.-Y.; Wu, H.H.; Tu, K.-H.; Tian, Y.-C. Incidence of Mortality, Acute Kidney Injury and Graft Loss in Adult Kidney Transplant Recipients with Coronavirus Disease 2019: Systematic Review and Meta-Analysis. *J. Clin. Med.* **2021**, *10*, 5162. https://doi.org/10.3390/jcm10215162

Academic Editor: Giorgio I. Russo

Received: 1 September 2021
Accepted: 28 October 2021
Published: 4 November 2021

Abstract: The adverse impact of Coronavirus disease 2019 (COVID-19) on kidney function has been reported since the global pandemic. The burden of COVID-19 on kidney transplant recipients, however, has not been systematically analyzed. A systematic review and meta-analysis with a random-effect model was conducted to explore the rate of mortality, intensive care unit admission, invasive mechanical ventilation, acute kidney injury, kidney replacement therapy and graft loss in the adult kidney transplant population with COVID-19. Sensitivity analysis, subgroup analysis and meta-regression were also performed. Results: we demonstrated a pooled mortality rate of 21% (95% CI: 19−23%), an intensive care unit admission rate of 26% (95% CI: 22–31%), an invasive ventilation rate among those who required intensive care unit care of 72% (95% CI: 62–81%), an acute kidney injury rate of 44% (95% CI: 39–49%), a kidney replacement therapy rate of 12% (95% CI: 9–15%), and a graft loss rate of 8% (95% CI: 5–15%) in kidney transplant recipients with COVID-19. The meta-regression indicated that advancing age is associated with higher mortality; every increase in age by 10 years was associated with an increased mortality rate of 3.7%. Regional differences in outcome were also detected. Further studies focused on treatments and risk factor identification are needed.

Keywords: acute kidney injury; coronavirus disease 2019; kidney replacement therapy; graft loss; mortality

1. Introduction

The novel coronavirus, severe acute respiratory syndrome coronavirus 2 (SARS-CoV-2), caused COVID-19 from the end of 2019 and has resulted in a huge burden on global healthcare systems. As of June 2021, more than 170 million people had been infected and nearly 4 million died. COVID-19 consists of a primary pulmonary infection with extensive systemic involvement. The overwhelming inflammatory response may lead to cytokine storm and multi-organ failure.

In severe COVID-19, dysregulated immunity induces endothelial injury, complement-mediated thrombosis and microangiopathy. The kidneys are one of the organs most involved during the progress of the disease; therefore, acute kidney injury (AKI) is common in patients with COVID-19 [1]. The incidence of AKI in COVID-19 varies across populations and critically ill patients seem to be the most susceptible [2,3]. In addition to being one of the negative impacts of COVID-19, AKI also serves as a predictor of mortality in patients with COVID-19 [4]. The kidney function reserve varies from patient to patient, depending on the kidneys' ability to handle external stresses or hazardous stimuli. In patients undergoing cardiac surgery, poorer preoperative kidney function is associated with a higher possibility of postoperative AKI [5]. Compared with the general population, kidney transplant recipients have a lower average kidney function reserve; thus, they are more susceptible to AKI [6,7].

Kidney transplant recipients are constantly at risk of complications associated with immunosuppression, which include opportunistic infections (e.g., BK virus, Epstein–Barr virus and cytomegalovirus infections), post-transplant lymphoproliferative disorder, and complications associated with immunosuppressants (e.g., calcineurin inhibitor-associated nephrotoxicity, calcineurin inhibitor- and corticosteroid-associated new-onset diabetes after transplantation and dyslipidemia) [8]. Because of their immunosuppressed status, the kidney transplant population is more susceptible to infection than the general population. The risk of COVID-19 transmission from a household contact is also higher in patients with solid organ or stem cell transplant [9]. In several studies, the reported outcomes of COVID-19 were worse in kidney transplant recipients with COVID-19 compared with the general population [10,11], although this has not been systematically analyzed. The aim of this study was to systematically review and analyze the outcomes of COVID-19 in kidney transplant recipients, including mortality rate, acute kidney injury rate, invasive ventilation rate and rate of graft loss.

2. Materials and Methods

2.1. Literature Search Strategy

We performed this meta-analysis in accordance with the preferred reporting items for systematic reviews and meta-analyses (PRISMA) guidelines for a meta-analysis of observational studies (Supplemental Table S1) [12]. We registered the protocol in PROSPERO (CRD42021260803). Two independent reviewers (J.J. Chen and G. Kuo) comprehensively searched for studies published before 08 June 2021 on PubMed, Medline, the Cochrane Library and Embase. The search strategy targeted published clinical trials, cohort studies, case series, letters to the editor and commentaries. The keyword and Mesh term used on PubMed were: (((COVID-19) OR (SARS-CoV-2)) OR (coronavirus)) AND ((((Kidney Transplantation) OR (Kidney transplant)) OR (Renal transplantation)) OR (Renal transplant)) with the following filters: Humans, Adult: 19+ years. English-language articles that were published from 2019–2021 were screened.

The other detailed search strategy and the results of that search process are provided in Supplemental Table S2. Review articles and meta-analyses were not included in our analysis, but their references were screened and searched for relevant studies.

2.2. Study Eligibility Criteria

The titles and abstracts of the studies returned by the search were examined by two reviewers (J.J. Chen and G. Kuo) independently, and articles were excluded upon initial screening of their titles or abstracts if these indicated that they were clearly irrelevant to the objective of the current study. The full texts of relevant articles were reviewed to determine whether the studies were eligible for inclusion. The inclusion criteria are: (i) studies enrolled an adult population with confirmed COVID-19, (ii) studies enrolled kidney transplant recipients, (iii) studies reported at least one of the outcomes of interest. The third reviewer (T.H. Lee) was consulted to reach an agreement through consensus in the case of any disagreement regarding eligibility. Studies were excluded if they were duplicated cohorts, presented insufficient information of outcomes, or included a pediatric population or a population of more than one organ transplantation.

2.3. Data Extraction and Outcome

Two investigators (J.J. Chen and G. Kuo) independently extracted the outcomes of interest and the characteristics of the included studies. The primary outcome in the present study was mortality in adult kidney transplant patients with COVID-19. The secondary outcomes included AKI, kidney replacement therapy (KRT), ICU admission, invasive mechanical ventilation (IMV), and graft loss. The IMV rate was calculated by the number of patients receiving IMV divided by the number of patients admitted to the ICU. The graft loss rate was calculated by the number of patients with graft loss divided by the number of patients who survived after COVID-19 infection.

2.4. Data Synthesis and Analysis

The analysis was conducted using the metaprop function in R package meta (version 4.18-2) [13]. The event rates of the outcomes of interest, including the mortality, ICU admission, AKI, IMV, KRT, and graft loss, were pooled and estimated. We chose the random-effect model because it is methodologically conservative for summary estimates and is more suitable as potential heterogeneity in the study populations may exist. Heterogeneity was examined by I^2 ($\geq$50% indicates substantial heterogeneity) and Cochran's Q statistic ($p < 0.1$ indicates moderate heterogeneity) [14–16]. The p values were two-sided, and statistical significance was set at $p < 0.05$. The sensitivity analysis was performed by removing the studies with a moderate-to-high risk of bias. The subgroup analysis was performed to explore potential sources of heterogeneity according to the study design (retrospective or prospective), location (single center or multi-center), patient number (<30 or at least 30), level of care (mixed in-patient and out-patient or purely in-patient), and areas of the countries where the study conducted (North America, Latin America, Europe, Asia). In the meta-regression, age and sample size were considered as potential mortality modifiers. Publication bias was assessed by the funnel plot and Egger's test; a p value of the Egger's test of < 0.1 indicates potential publication bias [16]. The statistical analysis was performed by using R software version 4.1.0 (The R Foundation, Vienna, Austria) [17].

2.5. Risk of Bias Assessment

The quality of the cohort studies was assessed independently by two authors (J.J. Chen and G. Kuo) using the Newcastle–Ottawa scale [18], which allocates a maximum of nine points for three major domains: quality of the selection, comparability, and outcome of study populations. Studies with a score of 7–9 were considered as low risk of bias, those with a score of 4–6 were considered as moderate risk of bias and those with a score of less than 4 were considered as high risk of bias. Disagreements between the two investigators (J.J. Chen and G. Kuo) were resolved by consensus with another author (T.H. Lee).

3. Results
3.1. Study Characteristics

The literature search flow is shown in Supplemental Figure S1. Through the electronic database search, there were 371 potentially eligible studies from PubMed, 496 potentially eligible studies from EMBASE, 385 studies from Medline, and 0 studies from the Cochrane review. After removing the duplicated articles, a total of 750 articles were screened according to their titles and abstracts. Seventy-eight full-text articles were further assessed for eligibility after screening (Supplemental Figure S1 & Supplemental Table S2). After excluding 19 studies for various reasons, 59 articles were included (Supplemental Table S3). A total of 59 studies comprising 5956 participants were enrolled in our study. Most of the studies were retrospective in design and enrolled hospitalized kidney transplant recipients with COVID-19. A minority of studies included a mixed level of care, where kidney transplant recipients with COVID-19 were managed either in the hospital or in an outpatient clinic (Table 1).

Table 1. Characteristics and outcomes of the included studies.

Study	Design	Country	Location	NOS	Age (Mean, y/o)	Sample Size	Mortality	ICU	MV	AKI	KRT	Graft Loss
Abolghasemi [19]	RC	Iran	Multi-center	6	49	24	10	12	NR	NR	NR	NR
Abrishami [20]	RC	Iran	Single-center	5	47.6	12	8	10	9	NR	NR	NR
Akalin [21]	RC	USA	Single-center	5	36	60	10	NR	11	NR	6	NR
Alberici [22]	RC	Italy	Single-center	5	59	20	5	4	0	6	1	NR
Azzi [23]	PC	USA	Single-center	5	59	229	47	NR	NR	NR	NR	NR
Azzi [23]	PC	USA	Single-center	5	61	79 ##	NR	NR	28	18	18	5
Banerjee [24]	RC	UK	Multi-center	6	57.4	7	1	4	2	4	3	NR
Bell [25]	RC	Scotland (Scottish Renal Registry)	Multi-center	7	NR	24	7	NR	NR	NR	NR	NR
Benotmane [26]	RC	France	Single-center	5	62.2	49	9	14	2	31	4	NR
Bossini [27]	RC	Italy	Multi-center	6	60	53	15	10	9	15	3	NR
Caillard [28]	RC	France (nationwide registry)	Multi-center	6	60.8	279	43	88	72	106	27	9
Chavarot [29]	RC	France	Multi-center	7	64.7	100	26	34	29	NR	NR	NR
Chen [30]	RC	USA	Single-center	5	56	30	6	NR	7	NR	4	NR
Coll [31]	RC	Spain	Multi-center	6	62	423	103	57	36	NR	NR	NR
Craig-Schapiro [32]	RC	USA, NY	Single-center	6	57	80	13	NR	16	25	4	4
Cravedi [33]	RC	USA	Multi-center	6	60	144	46	NR	42	74	NR	NR
Cristelli [34]	PC	Brazil	Single-center	5	53	491	140	NR	156	229	155	NR
Demir [35]	RC	Turkey	Multi-center	6	44.9	40	5	7	6	14	NR	NR
Devresse [36]	PC	USA	Single-center	5	57	22	2	2	2	5	0	NR
Dheir [37]	RC	Turkey	Single-center	5	48	20	2	NR	2	5	2	NR
Elec [38]	RC	Romania	Single-center	5	52	42	7	8	NR	10	N	NR
Elhadedy [39]	RC	UK	Multi-center	6	50.1	8	0	1	1	2	1	0
Elias [40]	PC	France	Multi-center	7	56.4	66	16	15	15	28	7	NR
Favà [41]	RC	Spain	Multi-center	6	59.7	104	28	NR	14	47	NR	NR
Fernández-Ruiz [42]	RC	Spain	Single-center	5	69.1	8	2	0	0	NR	NR	NR
Georgery [43]	RC	Belgium	Single-center	5	60.6	45	8	14	14	NR	NR	NR
Gupta [44]	PC	India	Single-center	5	44.5	10	1	NR	NR	NR	NR	NR
Hardesty [45]	RC	USA	Single-center	6	55	11	1	NR	3	1	1	NR
Hilbrands [46]	PC	Europe, ERACODA collaboration	Multi-center	7	60	305	65	57	49	30+	30	NR
Husain [47]	RC	USA, NY (Columbia University Vagelos College of Physicians and Surgeons)	Single-center	5	51	15	2	NR	4	6	NR	NR
Jager [48]	PC	Europe, ERA-EDTA Registry	Multi-center	7	60.9	1013	191	NR	NR	NR	NR	NR
Katz-Greenberg [49]	RC	USA, Philadelphia	Single-center	5	52.5	20	3	NR	4	9	1	1
Kumaresan [50]	RC	India	Single-center	5	49.7	16	3	NR	3	1	1	NR
Kute [51]	RC	India	Multi-center	6	43	250	29	53	30	121	24	12
Lubetzky [52]	RC	USA	Single-center	5	57	54	7	NR	11	21	3	3
Lum [53]	RC	USA	Single-center	5	48.5	41	4	9	8	11	4	1
Mamode [54]	RC	UK	Multi-center	5	56.2	121	36	30	22	NR	19	15
Maritati [55]	RC	Italy	Single-center	5	66	5	2	NR	3	1	1	NR
Meester [56]	PC	Belgium (NBVN Kidney Registry Group)	Multi-center	7	NR	43	6	NR	NR	NR	NR	NR
Mella [57]	RC	Italy	NR	5	55.5	6	4	NR	2	NR	NR	NR
Meziyerh [58]	RC	Netherlands	Single-center	5	56	15	6	6	5	NR	NR	NR
Mohamed [59]	PC	UK	Single-center	6	57	28	9	5	NR	14	2	NR
Molaei [60]	RC	Iran	Single-center	5	59.6	10	2	4	4	7	NR	NR

Table 1. *Cont.*

Study	Design	Country	Location	NOS	Age (Mean, y/o)	Sample Size	Mortality	ICU	MV	AKI	KRT	Graft Loss
Monfared [61]	RC	Iran	Single-center	5	52	22	6	NR	5	12	NR	NR
Montagud-Marrahi [62]	RC	Spain	Single-center	5	57.1	33	2	13	2	NR	NR	1
Nair [63]	RC	USA, NY	Single-center	5	57	10	3	5	4	5	1	NR
Oto [64]	RC	Turkey	Multi-center	6	48.4	109	14	22	19	46	4	NR
Ozturk [65]	RC	Turkey	Multi-center	7	48	81	9	17	14	NR	NR	NR
Phanish [66]	RC	UK	Multi-center	7	62	23	6	9	6	13	4	2
Pierrotti [67]	RC	Brazil	Single-center	5	51.9	51	13	23	17	30	19	0
Rodriguez-Cubillo [68]	RC	Spain	Single-center	5	66	24	6	NR	5	14	3	0
Sandes-Freitas [69]	RC	Brazil (National registry)	Multi-center	6	53.9	8	3	4	4	4	4	2
Santeusanio [70]	RC	USA	Single-center	5	43.8	38	11	NR	14	22	12	12
Shrivastava [71]	RC	USA	Single-center	5	61.5	38	9	13	9	27	6	NR
Tejada [72]	RC	USA, Detroit Medical Center	Single-center	5	56	25	1	4	1	16	NR	NR
Trujillo [73]	RC	Spain	Single-center	5	54	10	3	NR	0	8	NR	NR
Villa [74]	RC	Germany	Single-center	6	62	7	3	NR	3	4	NR	NR
Villanego [75]	PC	Spain (prospectively filled registry)	Multi-center	6	60	1011	220	140	NR	NR	NR	NR
Willicombe [76]	RC + PC [#]	UK, London	Single-center	5	56	113	17	NR	NR	NR	NR	NR
Zhu [77]	RC	China	Multi-center	6	45	10	1	NR	0	6	0	NR

[#] Including a seroprevalence survey and clinical cohort; [##] Sub-cohort report different outcomes other than mortality; abbreviations: AKI, Acute kidney injury; ICU, intensive care unit; KRT, kidney replacement therapy; MV, mechanical ventilation, NOS, Newcastle–Ottawa Scale; NR: not reported; PC, prospective cohort/case series; RC, retrospective cohort/case series.

3.2. Mortality of Adult Transplant Population with COVID-19 Infection

We found 58 studies with 5948 patients that reported the mortality of COVID-19 in adult kidney transplant recipients. The pooled mortality rate calculated by the random effect model was 21% (95% confidence interval [CI]: 19−23%) with moderate heterogeneity ($I^2 = 57\%$, $p < 0.01$) (Figure 1).

3.3. Secondary Outcomes of Adult Transplant Population with COVID-19 Infection

Thirty-two studies reported the rate of ICU admission. The pooled ICU admission rate was 26% (95% CI: 22−31%) with moderate heterogeneity ($I^2 = 80\%$, $p < 0.01$). The rates of IMV among patients requiring ICU care were reported in 21 studies. The pooled IMV rate was 72% (95% CI: 62–81%) with moderate heterogeneity ($I^2 = 65\%$, $p < 0.01$). The incidence rate of AKI was reported in 38 studies, with a pooled rate of 44% (95% CI: 39−49%) and moderate heterogeneity ($I^2 = 61\%$, $p < 0.01$). The requirement of KRT among patients experiencing AKI was reported in 27 studies. The pooled KRT rate among AKI patients was 30% (95% 22−39%) with high heterogeneity ($I^2 = 83\%$, $p < 0.01$), and was 12% (9−15%) with high heterogeneity ($I^2 = 83\%$, $p < 0.01$) among the whole-kidney transplant population. The rate of graft loss was reported in only 11 studies. The pooled graft loss rate among kidney transplant COVID-19 survivors was 8% (95% CI: 5−15%) with a high heterogeneity ($I^2 = 87\%$, $p < 0.01$). The summarized and detailed information of these secondary outcomes is depicted in Figure 2.

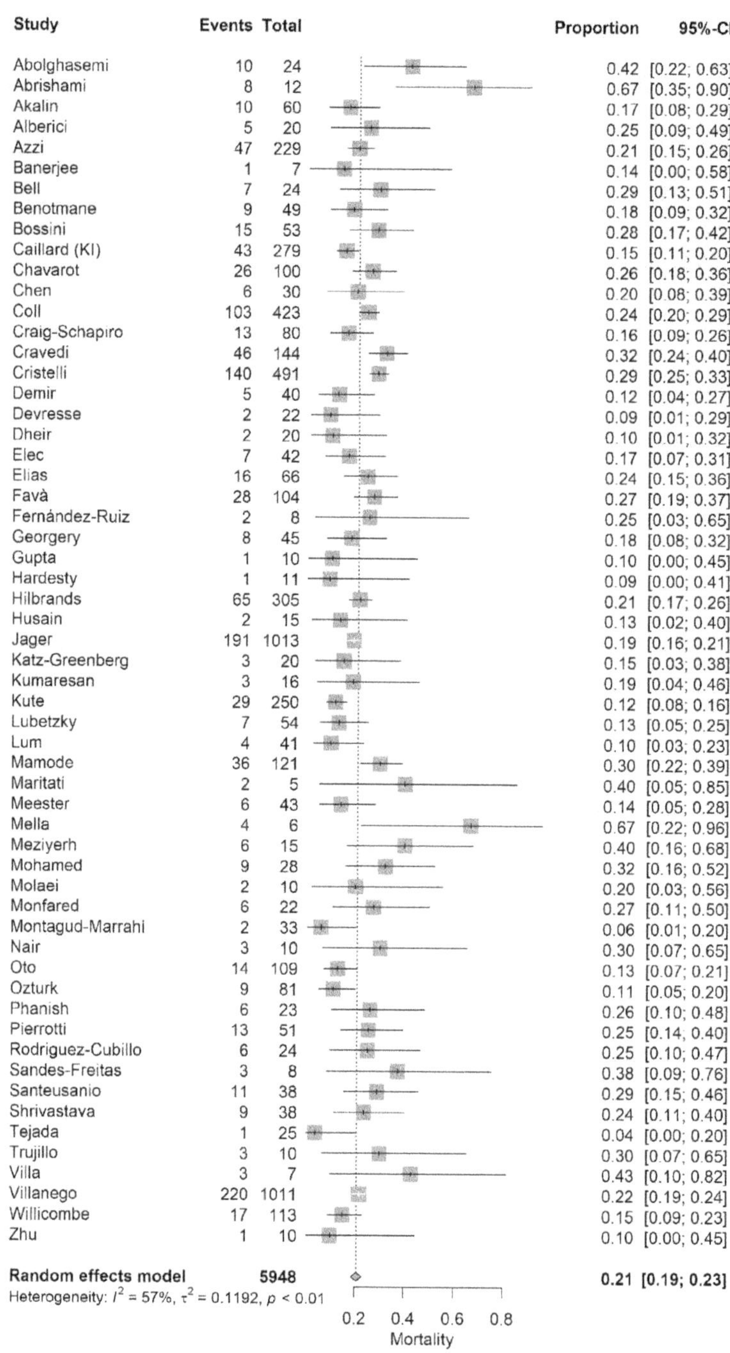

Figure 1. Forest plot of pooled incidence of mortality in the adult kidney transplant population with COVID-19 infection.

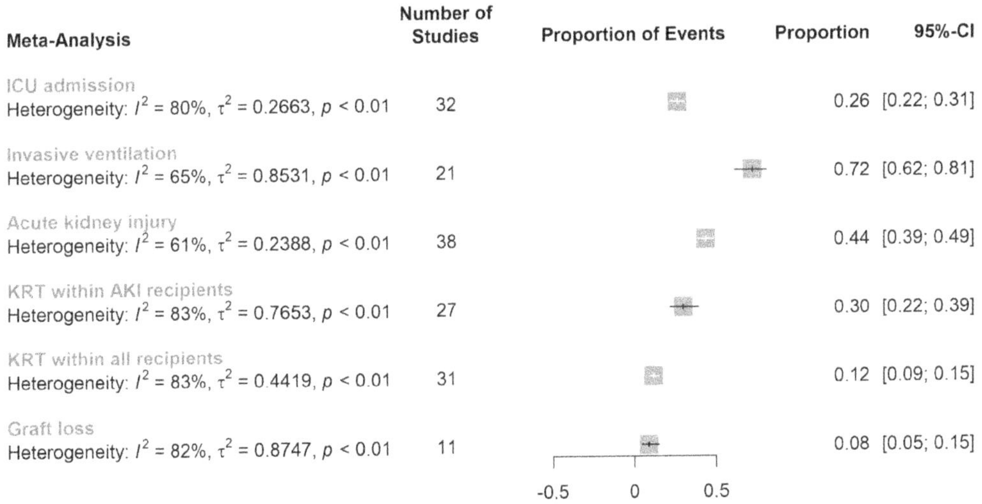

Meta-Analysis	Number of Studies	Proportion of Events	Proportion	95%-CI
ICU admission Heterogeneity: $I^2 = 80\%$, $\tau^2 = 0.2663$, $p < 0.01$	32		0.26	[0.22; 0.31]
Invasive ventilation Heterogeneity: $I^2 = 65\%$, $\tau^2 = 0.8531$, $p < 0.01$	21		0.72	[0.62; 0.81]
Acute kidney injury Heterogeneity: $I^2 = 61\%$, $\tau^2 = 0.2388$, $p < 0.01$	38		0.44	[0.39; 0.49]
KRT within AKI recipients Heterogeneity: $I^2 = 83\%$, $\tau^2 = 0.7653$, $p < 0.01$	27		0.30	[0.22; 0.39]
KRT within all recipients Heterogeneity: $I^2 = 83\%$, $\tau^2 = 0.4419$, $p < 0.01$	31		0.12	[0.09; 0.15]
Graft loss Heterogeneity: $I^2 = 82\%$, $\tau^2 = 0.8747$, $p < 0.01$	11		0.08	[0.05; 0.15]
		-0.5 0 0.5		

Figure 2. Forest plot of secondary outcome in the adult kidney transplant population with COVID-19 infection.

3.4. Publication and Risk of Bias of Enrolled Studies

Potential publication bias was illustrated by a funnel plot (Supplemental Figure S2). The Egger's test of funnel asymmetry displayed an insignificant result (p value = 0.64). The risk of bias was assessed via the Newcastle–Ottawa scale for non-randomized studies; we summarize the risk of bias of each study in Supplemental Table S4. For retrospective studies with a single-center population, we considered that there was a potential risk regarding the representativeness of the exposed cohort. Owing to the retrospective study design in nature, most of the enrolled studies were without control group and therefore the domain score regarding the comparability of cohorts was considered as zero.

3.5. Sensitivity Analysis, Subgroup Analysis and Meta-Regression

The sensitivity analysis was performed by including only studies with a low risk of bias (NOS scale higher or equal to 7). A total of eight studies fulfilled this criterion. The pooled mortality was 20% (95% CI: 18−22%) with low heterogeneity ($I^2 = 35\%$, $p = 0.15$). (Figure 3A).

We performed a subgroup analysis by dividing the studies into different groups: (1) study design, (2) single- or multi-center, (3) patient number, (4) level of care, and (5) areas of study countries. There were no interactions for study design, single or multi-center, patient numbers, or level of care. There was significant interaction between the different areas of study, with a higher mortality rate reported for Latin America. ($p < 0.01$) (Figure 3B).

A meta-regression was performed to examine the relationship between two covariates (sample size and age) and mortality. Sample size was not significantly associated with increased mortality risk (Figure 4A). The mortality rate was significantly associated with older age (mortality rate = 0.0037 × Age − 0.0092, p value = 0.023). This implies a 3.7% increase in mortality rate for every 10 years of age (Figure 4B).

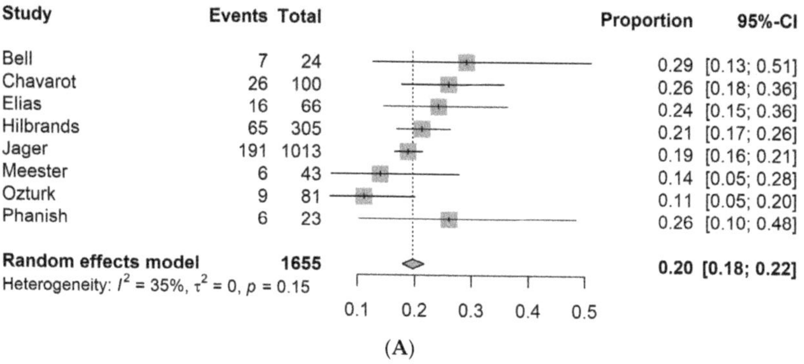

Figure 3. Sensitivity analysis after excluding studies with moderate-to-high risk of bias ((**A**), upper) and forest plot of subgroup analysis ((**B**), lower).

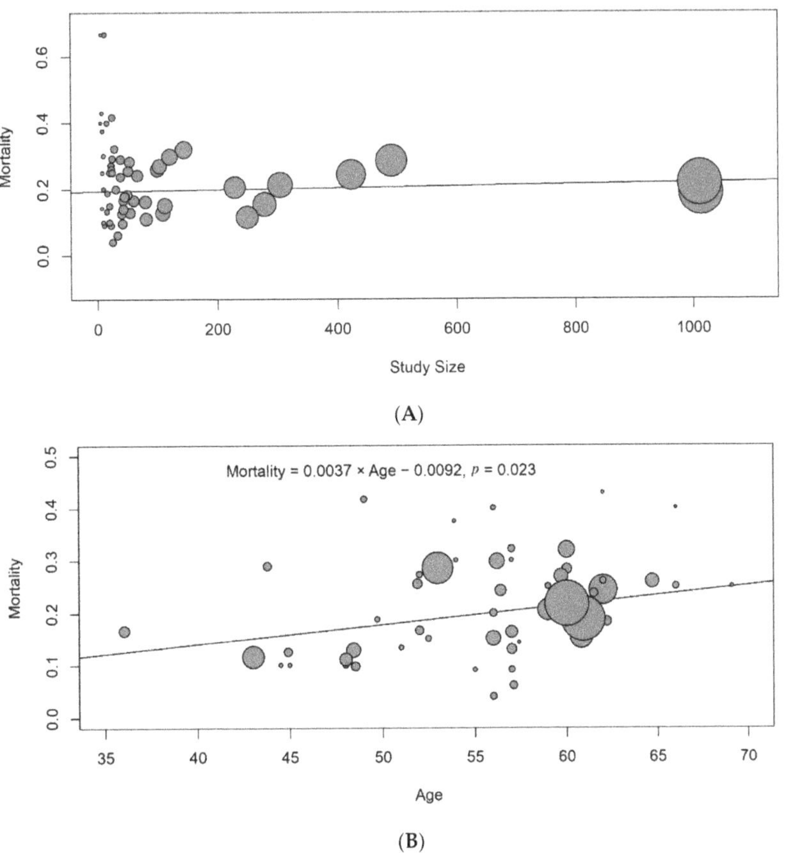

Figure 4. Meta-regression according to sample size ((**A**), upper) and age ((**B**), lower).

4. Discussion

In the present systematic review and meta-analysis, two points are worth noting. First, we demonstrated higher rates of adverse outcomes in adult kidney transplant recipients with COVID-19, and these included mortality (21%), ICU admission (26%), IMV among those who required ICU care (72%), AKI (44%), KRT (12%) and graft loss (8%). Second, older age is associated with an increase in mortality rate.

Compared to the general population, the incidence of AKI and KRT and the mortality rate in kidney transplant recipients are higher [22,24,78,79]. In the non-transplant population, Meyerowitz-Katz et al. reported a pooled infection fatality rate of 0.68% among COVID-19 patients [78]. Levin et al. reported that the fatality rate was increased with advanced age. The age-specific fatality rate of COVID-19 increased from less than 0.4% below age 55 to 15% at age 85 [80]. In kidney transplant recipients, the excessive deaths may reflect their immunocompromised status and susceptibility to infection. Using the registry data of Australia and New Zealand, Chan et al. demonstrated a higher infection-related mortality rate in kidney transplant recipients compared to the general population. Ozturk et al. compared the clinical characteristics of COVID-19 among kidney transplant recipients, HD, CKD and control groups (patients without kidney diseases) [65]. Overall, the kidney transplant recipients demonstrated a higher mortality risk than the control group, but were still less vulnerable to death than CKD and HD patients [65]. In addition to the severity of COVID-19 per se, co-infection with other viruses can increase the risk of disease progression and mortality in kidney transplant recipients. Molaei et al. reported that co-infection of COVID-19 and cytomegalovirus, the most prevalent pathogen in kidney transplant recipients, might increase mortality risk [60]. In addition, potential drug–drug interactions between immunosuppressive medications and antiviral therapy could complicate and worsen the clinical condition [81]. In patients who recovered from COVID-19, those with immunocompromised status may still experience longer viral shedding from the respiratory tract than the general population [82]. The increased susceptibility to infection and delayed viral clearance both indicate that immunocompromised status contributes to higher mortality in kidney transplant recipients. These findings may explain the higher mortality risk among kidney transplant recipients when compared to the general population.

In the present study, we demonstrated that older age is a risk factor of mortality in kidney transplant recipients with COVID-19. The observation of age-related mortality is in agreement with the study by Chan et al. [83]. This is not surprising, because age has been shown to be a crucial risk factor for mortality in different populations [46,84–86].

AKI is common during the routine care of kidney transplant recipients. In a retrospective longitudinal cohort study using the US Renal Data System, 11.6% of kidney transplant recipients experienced episodes of AKI, and 14.8% of those who experienced AKI required temporary KRT [6]. This higher susceptibility to AKI is also observed in COVID-19 patients. According to the meta-analysis published by Chen et al., the pooled AKI occurrence in the general population was 8.9% [79]. Among the studies enrolled in the meta-analysis by Chen et al., there were two studies reporting the AKI rates of kidney transplant recipients with COVID-19, which were 30–57% [22,24]. The use of calcineurin inhibitors by kidney transplant recipients can cause vascular and endothelial damage. In COVID-19, vessels and endothelium are targets for viral attack. The baseline susceptibility to endothelial injury by an external insult to the vasculature may make kidney transplant recipients more vulnerable to microvascular injury and thrombosis [87–89]. In our study, the pooled AKI rate was 43% in the kidney transplant recipients, which is strikingly higher than in the general population. Among those patients with AKI, 30% required KRT. The KRT rate was higher than in previous reports on the general population, and this indicates that COVID-19 might pose a higher risk of AKI and severe AKI requiring dialysis in kidney transplant recipients than in the general population.

In this study, the rate of ICU admission (26%) and IMV rate (73%) were similar to the data from the general population. In a meta-analysis, Chang et al. included 12,437 ICU-admitted COVID-19 patients, 69% of whom required IMV [90]. However, the actual rate of ICU admission and mechanical ventilation may be difficult to compare directly between studies because the disease prevalence may vary between different times and countries. The capacity of ICUs and their criteria of admission may also differ among healthcare systems.

In the present study, we also observed a difference in mortality across different areas. Another meta-analysis demonstrated that the mortality of COVID-19 patients with at least one comorbidity is higher in Europe and Latin America [91]. This observation, however, should be interpreted with caution. The differences between healthcare systems and the timing and variants of the SARS-CoV-2 pandemic, the implementation of non-pharmaceutical intervention against viral spreading, and the speed of vaccination could lead to variability in the extent and severity of COVID-19. The relationship between mortality and geographic region requires further observation and investigation through larger-scale, multinational studies.

This study features some limitations. First, information on the detailed immunosuppressive regimens, the concentration of immunosuppression medications, the type of the transplants (deceased or living) and the induction therapy regimens are lacking in most studies. Therefore, we could not adjust the outcomes with these factors. Among the studies with smaller patient numbers, the majority of baseline immunosuppressive agents were similar to those used in our daily practice, which included a combination of prednisolone, mycophenolate mofetil/mycophenolic acid, and tacrolimus. A smaller fraction of patients receive cyclosporine instead of tacrolimus, and a much smaller proportion of patients take leflunomide or azathioprine. The adjustment of immunosuppressants took the form of either a reduction in the dosage or temporary discontinuation. Basuki et al. reported a lower mortality with cyclosporine in kidney transplant recipients than with other immunosuppressive drugs during the treatment of COVID-19 [92]. However, only small and limited studies reported the differences between immunosuppressive regimens. Larger registries contain no detailed data for the comparison. Second, most studies do not report the severity of patients' symptoms by any known critical care scoring systems; therefore, we could not classify patients' disease severity beyond dividing them into an inpatient population and a population of patients managed on an outpatient basis. In addition, advancements in COVID-19 knowledge and management might result in improvements in COVID-19 prognosis. However, owing to this, information on COVID-19 treatment strategies is lacking in most studies. We analyzed the possible impact of earlier and recent recruitment of patients on clinical outcomes using subgroup analysis and meta-regression by stratifying the studies according to the date of final assessment or enrollment before or after July 2020. Using these two analysis models, we did not observe any significant difference in mortality between early and recent studies (data not shown). Furthermore, most studies are retrospective in design and based on a single -enter population. Although we performed a meta-regression examining the relationship between sample size and mortality, the risk of bias regarding the lack of representativeness and adequate comparability should be noted.

5. Conclusions

Our study demonstrated that adult kidney transplant recipients with COVID-19 had a high rate of mortality, AKI, and KRT. The risk of mortality increased in proportion with the recipients' age. Further studies focused on risk factor stratification, immunosuppressant drugs adjustment, and antiviral treatment in kidney transplant populations with COVID-19 infection are needed.

Supplementary Materials: The following are available online at https://www.mdpi.com/article/10.3390/jcm10215162/s1. Figure S1: PRISMA Flow Diagram. Figure S2: Funnel plot for publication bias assessment. Table S1: PRISMA checklist. Table S2: Details of Search Strategy results for each database. Table S3: Primary reasons for exclusion of excluded studies; Table S4: Newcastle–Ottawa Scale assessment of the included studies.

Author Contributions: Y.-C.T. and J.-J.C. participated in the conceptualization (creating the ideas and overarching research goal), J.-J.C. and G.K. participated in the methodology, validation and writing—original draft, G.K. carried out the software programing and supporting algorithms, G.K. and J.-J.C. carried out the formal analysis, T.H.L., J.-J.C. and G.K. carried out the investigation and data curation, H.-Y.Y. participated in resourcing and visualization, Y.-C.T., H.H.W. and K.-H.T. carried out the writing—review and editing, Y.-C.T., H.-Y.Y. carried out the supervision and project administration. All authors have read and agreed to the published version of the manuscript.

Funding: This work was supported by grants from Chang Gung Memorial Hospital Project to Ya-Chung Tian (CORPG3K0281).

Institutional Review Board Statement: Not applicable.

Informed Consent Statement: Not applicable.

Data Availability Statement: Not applicable.

Conflicts of Interest: The authors declare no conflict of interest.

References

1. Perico, L.; Benigni, A.; Casiraghi, F.; Ng, L.F.P.; Renia, L.; Remuzzi, G. Immunity, endothelial injury and complement-induced coagulopathy in COVID-19. *Nat. Rev. Nephrol.* **2021**, *17*, 46–64. [CrossRef] [PubMed]
2. Argenziano, M.G.; Bruce, S.L.; Slater, C.L.; Tiao, J.R.; Baldwin, M.R.; Barr, R.G.; Chang, B.P.; Chau, K.H.; Choi, J.J.; Gavin, N.; et al. Characterization and clinical course of 1000 patients with coronavirus disease 2019 in New York: Retrospective case series. *BMJ* **2020**, *369*, m1996. [CrossRef] [PubMed]

3.   Hirsch, J.S.; Ng, J.H.; Ross, D.W.; Sharma, P.; Shah, H.H.; Barnett, R.L.; Hazzan, A.D.; Fishbane, S.; Jhaveri, K.D.; Abate, M.; et al. Acute kidney injury in patients hospitalized with COVID-19. *Kidney Int.* **2020**, *98*, 209–218. [CrossRef] [PubMed]
4.   Gupta, S.; Hayek, S.S.; Wang, W.; Chan, L.; Mathews, K.S.; Melamed, M.L.; Brenner, S.K.; Leonberg-Yoo, A.; Schenck, E.J.; Radbel, J.; et al. Factors associated with death in critically ill patients with coronavirus disease 2019 in the US. *JAMA Intern. Med.* **2020**, *180*, 1436–1447. [CrossRef]
5.   Husain-Syed, F.; Ferrari, F.; Sharma, A.; Danesi, T.H.; Bezerra, P.; Lopez-Giacoman, S.; Samoni, S.; de Cal, M.; Corradi, V.; Virzì, G.M.; et al. Preoperative renal functional reserve predicts risk of acute kidney injury after cardiac operation. *Ann. Thorac. Surg.* **2018**, *105*, 1094–1101. [CrossRef]
6.   Mehrotra, A.; Rose, C.; Pannu, N.; Gill, J.; Tonelli, M.; Gill, J.S. Incidence and consequences of acute kidney injury in kidney transplant recipients. *Am. J. Kidney Dis.* **2012**, *59*, 558–565. [CrossRef]
7.   Hundemer, G.L.; Srivastava, A.; A Jacob, K.; Krishnasamudram, N.; Ahmed, S.; Boerger, E.; Sharma, S.; Pokharel, K.K.; A Hirji, S.; Pelletier, M.; et al. Acute kidney injury in renal transplant recipients undergoing cardiac surgery. *Nephrol. Dial. Transplant.* **2021**, *36*, 185–196. [CrossRef]
8.   Yeo, W.-S.; Ng, Q.X. Biomarkers of immune tolerance in kidney transplantation: An overview. *Pediatr. Nephrol.* **2021**, 1–10. [CrossRef]
9.   Lewis, N.M.; Chu, V.T.; Ye, D.; E Conners, E.; Gharpure, R.; Laws, R.L.; E Reses, H.; Freeman, B.D.; Fajans, M.; Rabold, E.M.; et al. Household transmission of severe acute respiratory syndrome coronavirus-2 in the United States. *Clin. Infect. Dis.* **2021**, *73*, e1805–e1813. [CrossRef]
10.   Marinaki, S.; Tsiakas, S.; Korogiannou, M.; Grigorakos, K.; Papalois, V.; Boletis, I. A systematic review of COVID-19 infection in kidney transplant recipients: A universal effort to preserve patients' lives and allografts. *J. Clin. Med.* **2020**, *9*, 2986. [CrossRef]
11.   Choi, M.; Bachmann, F.; Naik, M.G.; Duettmann, W.; Duerr, M.; Zukunft, B.; Schwarz, T.; Corman, V.M.; Liefeldt, L.; Budde, K.; et al. Low seroprevalence of SARS-CoV-2 antibodies during systematic antibody screening and serum responses in patients after COVID-19 in a German transplant center. *J. Clin. Med.* **2020**, *9*, 3401. [CrossRef]
12.   Moher, D.; Liberati, A.; Tetzlaff, J.; Altman, D.G. Preferred reporting items for systematic reviews and meta-analyses: The PRISMA statement. *BMJ* **2009**, *339*, b2535. [CrossRef]
13.   Viechtbauer, W. Conducting meta-analyses in R with the metafor package. *J. Stat. Softw.* **2010**, *36*, 1–48. [CrossRef]
14.   Translator Higgins, J.P.T.; Thompson, S.G.; Deeks, J.J.; Altman, D.G. Measuring inconsistency in meta-analyses. *Br. Med. J.* **2003**, *327*, 557–560. [CrossRef]
15.   Higgins, J.P.T.; Thompson, S.G. Quantifying heterogeneity in a meta-analysis. *Stat. Med.* **2002**, *21*, 1539–1558. [CrossRef]
16.   Higgins, J.P.; Thomas, J.; Chandler, J.; Cumpston, M.; Li, T.; Page, M.J.; Welch, V.A. *Cochrane Handbook for Systematic Reviews of Interventions*, 6th ed.; Cochrane: Oxford, UK, 2021; Available online: www.training.cochrane.org/handbook (accessed on 15 March 2021).
17.   Schoch, D.; Sommer, R.; Augustin, M.; Ständer, S.; Blome, C. Patient-reported outcome measures in pruritus: A systematic review of measurement properties. *J. Investig. Dermatol.* **2017**, *137*, 2069–2077. [CrossRef]
18.   Wells, G.A.; Shea, B.; O'Connell, D.; Peterson, J.; Welch, V.; Losos, M.; Tugwell, P. The Newcastle-Ottawa Scale (NOS) for Assessing the Quality of Nonrandomised Studies in Meta-Analyses. 2013. Available online: http://www.ohri.ca/programs/clinicalepidemiology/oxford.asp (accessed on 15 March 2021).
19.   Abolghasemi, S.; Mardani, M.; Sali, S.; Honarvar, N.; Baziboroun, M. COVID-19 and kidney transplant recipients. *Transpl. Infect. Dis.* **2020**, *22*, e13413. [CrossRef]
20.   Abrishami, A.; Samavat, S.; Behnam, B.; Arab-Ahmadi, M.; Nafar, M.; Taheri, M.S. Clinical course, imaging features, and outcomes of COVID-19 in kidney transplant recipients. *Eur. Urol.* **2020**, *78*, 281–286. [CrossRef]
21.   Akalin, E.; Azzi, Y.; Bartash, R.; Seethamraju, H.; Parides, M.; Hemmige, V.; Ross, M.; Forest, S.; Goldstein, D.Y.; Ajaimy, M.; et al. Covid-19 and kidney transplantation. *N. Engl. J. Med.* **2020**, *382*, 2475–2477. [CrossRef]
22.   Alberici, F.; Delbarba, E.; Manenti, C.; Econimo, L.; Valerio, F.; Pola, A.; Maffei, C.; Possenti, S.; Zambetti, N.; Moscato, M.; et al. A single center observational study of the clinical characteristics and short-term outcome of 20 kidney transplant patients admitted for SARS-CoV2 pneumonia. *Kidney Int.* **2020**, *97*, 1083–1088. [CrossRef]
23.   Azzi, Y.; Parides, M.; Alani, O.; Loarte-Campos, P.; Bartash, R.; Forest, S.; Colovai, A.; Ajaimy, M.; Liriano-Ward, L.; Pynadath, C.; et al. COVID-19 infection in kidney transplant recipients at the epicenter of pandemics. *Kidney Int.* **2020**, *98*, 1559–1567. [CrossRef]
24.   Banerjee, D.; Popoola, J.; Shah, S.; Ster, I.C.; Quan, V.; Phanish, M. COVID-19 infection in kidney transplant recipients. *Kidney Int.* **2020**, *97*, 1076–1082. [CrossRef]
25.   Bell, S.; Campbell, J.; McDonald, J.; O'Neill, M.; Watters, C.; Buck, K.; Cousland, Z.; Findlay, M.; Lone, N.I.; Metcalfe, W.; et al. COVID-19 in patients undergoing chronic kidney replacement therapy and kidney transplant recipients in Scotland: Findings and experience from the Scottish renal registry. *BMC Nephrol.* **2020**, *21*, 419. [CrossRef]
26.   Benotmane, I.; Perrin, P.; Vargas, G.G.; Bassand, X.; Keller, N.; Lavaux, T.; Ohana, M.; Bedo, D.; Baldacini, C.; Sagnard, M.; et al. Biomarkers of cytokine release syndrome predict disease severity and mortality from COVID-19 in kidney transplant recipients. *Transplantation* **2021**, *105*, 158–169. [CrossRef]

27. Bossini, N.; Alberici, F.; Delbarba, E.; Valerio, F.; Manenti, C.; Possenti, S.; Econimo, L.; Maffei, C.; Pola, A.; Terlizzi, V.; et al. Kidney transplant patients with SARS-CoV-2 infection: The Brescia Renal COVID task force experience. *Arab. Archaeol. Epigr.* **2020**, *20*, 3019–3029. [CrossRef]

28. Caillard, S.; Anglicheau, D.; Matignon, M.; Durrbach, A.; Greze, C.; Frimat, L.; Thaunat, O.; Legris, T.; Moal, V.; Westeel, P.F.; et al. An initial report from the French SOT COVID Registry suggests high mortality due to COVID-19 in recipients of kidney transplants. *Kidney Int.* **2020**, *98*, 1549–1558. [CrossRef]

29. Chavarot, N.; Gueguen, J.; Bonnet, G.; Jdidou, M.; Trimaille, A.; Burger, C.; Amrouche, L.; Weizman, O.; Pommier, T.; Aubert, O.; et al. COVID-19 severity in kidney transplant recipients is similar to nontransplant patients with similar comorbidities. *Arab. Archaeol. Epigr.* **2021**, *21*, 1285–1294. [CrossRef]

30. Chen, T.Y.; Farghaly, S.; Cham, S.; Tatem, L.L.; Sin, J.H.; Rauda, R.; Ribisi, M.; Sumrani, N. COVID-19 pneumonia in kidney transplant recipients: Focus on immunosuppression management. *Transpl. Infect. Dis.* **2020**, *22*, e13378. [CrossRef] [PubMed]

31. Coll, E.; Fernández-Ruiz, M.; Sánchez-Álvarez, J.E.; Martínez-Fernández, J.R.; Crespo, M.; Gayoso, J.; Bada-Bosch, T.; Oppenheimer, F.; Moreso, F.; López-Oliva, M.O.; et al. COVID-19 in transplant recipients: The Spanish experience. *Am. J. Transplant.* **2021**, *21*, 1825–1837. [CrossRef] [PubMed]

32. Craig-Schapiro, R.; Salinas, T.; Lubetzky, M.; Abel, B.T.; Sultan, S.; Lee, J.R.; Kapur, S.; Aull, M.J.; Dadhania, D.M. COVID-19 outcomes in patients waitlisted for kidney transplantation and kidney transplant recipients. *Arab. Archaeol. Epigr.* **2021**, *21*, 1576–1585. [CrossRef] [PubMed]

33. Cravedi, P.; Mothi, S.S.; Azzi, Y.; Haverly, M.; Farouk, S.S.; Pérez-Sáez, M.J.; Redondo-Pachón, M.D.; Murphy, B.; Florman, S.; Cyrino, L.G.; et al. COVID-19 and kidney transplantation: Results from the TANGO International Transplant Consortium. *Arab. Archaeol. Epigr.* **2020**, *20*, 3140–3148. [CrossRef]

34. Cristelli, M.P.; Viana, L.A.; Dantas, M.T.; Martins, S.B.; Fernandes, R.; Nakamura, M.R.; Santos, D.W.; Taddeo, J.B.; Azevedo, V.F.; Foresto, R.D.; et al. The full spectrum of COVID-19 development and recovery among kidney transplant recipients. *Transplantation* **2021**, *105*, 1433–1444. [CrossRef]

35. Demir, E.; Uyar, M.; Parmaksiz, E.; Sinangil, A.; Yelken, B.; Dirim, A.B.; Merhametsiz, O.; Yadigar, S.; Ucar, Z.A.; Ucar, A.R.; et al. COVID-19 in kidney transplant recipients: A multicenter experience in Istanbul. *Transpl. Infect. Dis.* **2020**, *22*, e13371. [CrossRef]

36. Devresse, A.; Belkhir, L.; Vo, B.; Ghaye, B.; Scohy, A.; Kabamba, B.; Goffin, E.; De Greef, J.; Mourad, M.; De Meyer, M.; et al. COVID-19 infection in kidney transplant recipients: A single-center case series of 22 cases from Belgium. *Kidney Med.* **2020**, *2*, 459–466. [CrossRef]

37. Dheir, H.; SİPAHİ, S.; Yaylaci, S.; Çetin, E.S.; Genç, A.B.; Firat, N.; Köroğlu, M.; Muratdaği, G.; Tomak, Y.; Özmen, K.; et al. Clinical course of COVID-19 disease in immunosuppressed renal transplant patients. *Turk. J. Med. Sci.* **2021**, *51*, 428–434. [CrossRef]

38. Elec, A.D.; Oltean, M.; Goldis, P.; Cismaru, C.; Lupse, M.; Muntean, A.; Elec, F.I. COVID-19 after kidney transplantation: Early outcomes and renal function following antiviral treatment. *Int. J. Infect. Dis.* **2021**, *104*, 426–432. [CrossRef]

39. Elhadedy, M.A.; Marie, Y.; Halawa, A. COVID-19 in renal transplant recipients: Case series and a brief review of current evidence. *Nephron* **2020**, *145*, 1–7. [CrossRef]

40. Elias, M.; Pievani, D.; Randoux, C.; Louis, K.; Denis, B.; DeLion, A.; Le Goff, O.; Antoine, C.; Greze, C.; Pillebout, E.; et al. COVID-19 infection in kidney transplant recipients: Disease incidence and clinical outcomes. *J. Am. Soc. Nephrol.* **2020**, *31*, 2413–2423. [CrossRef]

41. Favà, A.; Cucchiari, D.; Montero, N.; Toapanta, N.; Centellas, F.J.; Vila-Santandreu, A.; Coloma, A.; Meneghini, M.; Manonelles, A.; Sellarés, J.; et al. Clinical characteristics and risk factors for severe COVID-19 in hospitalized kidney transplant recipients: A multicentric cohort study. *Arab. Archaeol. Epigr.* **2020**, *20*, 3030–3041. [CrossRef]

42. Fernández-Ruiz, M.; Andrés, A.; Loinaz, C.; Delgado, J.F.; López-Medrano, F.; Juan, R.S.; González, E.; Polanco, N.; Folgueira, M.D.; Lalueza, A.; et al. COVID-19 in solid organ transplant recipients: A single-center case series from Spain. *Arab. Archaeol. Epigr.* **2020**, *20*, 1849–1858. [CrossRef]

43. Georgery, H.; Devresse, A.; Scohy, A.; Kabamba, B.; Darius, T.; Buemi, A.; De Greef, J.; Belkhir, L.; Yombi, J.-C.; Goffin, E.; et al. The second wave of COVID-19 disease in a kidney transplant recipient cohort: A single-center experience in Belgium. *Transplantation* **2021**, *105*, e41–e42. [CrossRef]

44. Gupta, A.; Kute, V.B.; Patel, H.V.; Engineer, D.P.; Banerjee, S.; Modi, P.R.; Rizvi, S.J.; Mishra, V.V.; Patel, A.H.; Navadiya, V. Feasibility of convalescent plasma therapy in kidney transplant recipients with severe COVID-19: A single-center prospective cohort study. *Exp. Clin. Transplant.* **2021**, *19*, 304–309. [CrossRef] [PubMed]

45. Hardesty, A.; Pandita, A.; Vieira, K.; Rogers, R.; Merhi, B.; Osband, A.J.; Aridi, J.; Shi, Y.; Bayliss, G.; Cosgrove, C.; et al. Coronavirus disease 2019 in kidney transplant recipients: Single-center experience and case-control study. *Transplant. Proc.* **2021**, *53*, 1187–1193. [CrossRef]

46. Hilbrands, L.B.; Duivenvoorden, R.; Vart, P.; Franssen, C.F.M.; Hemmelder, M.H.; Jager, K.J.; Kieneker, L.M.; Noordzij, M.; Pena, M.J.; de Vries, H.; et al. COVID-19-related mortality in kidney transplant and dialysis patients: Results of the ERACODA collaboration. *Nephrol. Dial. Transplant.* **2020**, *35*, 1973–1983. [CrossRef] [PubMed]

47. The Columbia University Kidney Transplant Program. Early description of coronavirus 2019 disease in kidney transplant recipients in New York. *J. Am. Soc. Nephrol.* **2020**, *31*, 1150–1156. [CrossRef] [PubMed]

48. Jager, K.J.; Kramer, A.; Chesnaye, N.C.; Couchoud, C.; Sánchez-Álvarez, J.E.; Garneata, L.; Collart, F.; Hemmelder, M.H.; Ambühl, P.; Kerschbaum, J.; et al. Results from the ERA-EDTA Registry indicate a high mortality due to COVID-19 in dialysis patients and kidney transplant recipients across Europe. *Kidney Int.* **2020**, *98*, 1540–1548. [CrossRef]

49. Katz-Greenberg, G.; Yadav, A.; Gupta, M.; Martinez-Cantarin, M.P.; Gulati, R.; Ackerman, L.; Belden, K.; Singh, P. Outcomes of COVID-19-positive kidney transplant recipients: A single-center experience. *Clin. Nephrol.* **2020**, *94*, 318–321. [CrossRef] [PubMed]

50. Kumaresan, M.; Babu, M.; Parthasarathy, R.; Matthew, M.; Kathir, C.; Rohit, A. Clinical profile of SARS-CoV-2 infection in kidney transplant patients-A single centre observational study. *Indian J. Transplant.* **2020**, *14*, 288. [CrossRef]

51. Kute, V.B.; Bhalla, A.K.; Guleria, S.; Ray, D.S.; Bahadur, M.M.; Shingare, A.; Hegde, U.; Gang, S.; Raju, S.; Patel, H.V.; et al. Clinical profile and outcome of COVID-19 in 250 kidney transplant recipients: A multicenter cohort study from India. *Transplantation* **2021**, *105*, 851–860. [CrossRef]

52. Lubetzky, M.; Aull, M.J.; Craig-Schapiro, R.; Lee, J.R.; Marku-Podvorica, J.; Salinas, T.; Gingras, L.; Lee, J.B.; Sultan, S.; Kodiyanplakkal, R.P.; et al. Kidney allograft recipients, immunosuppression, and coronavirus disease-2019: A report of consecutive cases from a New York City transplant center. *Nephrol. Dial. Transplant.* **2020**, *35*, 1250–1261. [CrossRef]

53. Lum, E.; Bunnapradist, S.; Multani, A.; Beaird, O.E.; Carlson, M.; Gaynor, P.; Kotton, C.; Abdalla, B.; Danovitch, G.; Kendrick, E.; et al. Spectrum of coronavirus disease 2019 outcomes in kidney transplant recipients: A single-center experience. *Transplant. Proc.* **2020**, *52*, 2654–2658. [CrossRef]

54. Mamode, N.; Ahmed, Z.; Jones, G.; Banga, N.; Motallebzadeh, R.; Tolley, H.; Marks, S.; Stojanovic, J.; Khurram, M.A.; Thuraisingham, R.; et al. Mortality rates in transplant recipients and transplantation candidates in a high-prevalence COVID-19 environment. *Transplantation* **2021**, *105*, 212–215. [CrossRef]

55. Maritati, F.; Cerutti, E.; Zuccatosta, L.; Fiorentini, A.; Finale, C.; Ficosecco, M.; Cristiano, F.; Capestro, A.; Balestra, E.; Taruscia, D.; et al. SARS-CoV-2 infection in kidney transplant recipients: Experience of the Italian Marche region. *Transpl. Infect. Dis.* **2020**, *22*, e13377. [CrossRef]

56. De Meester, J.; De Bacquer, D.; Naesens, M.; Meijers, B.; Couttenye, M.M.; De Vriese, A.S.; For the NBVN Kidney Registry Group. Incidence, characteristics, and outcome of COVID-19 in adults on kidney replacement therapy: A regionwide registry study. *J. Am. Soc. Nephrol.* **2021**, *32*, 385–396. [CrossRef] [PubMed]

57. Mella, A.; Mingozzi, S.; Gallo, E.; Lavacca, A.; Rossetti, M.; Clari, R.; Randone, O.; Maffei, S.; Salomone, M.; Imperiale, D.; et al. Case series of six kidney transplanted patients with COVID-19 pneumonia treated with tocilizumab. *Transpl. Infect. Dis.* **2020**, *22*, e13348. [CrossRef]

58. Meziyerh, S.; Van Der Helm, D.; de Vries, A. Vulnerabilities in kidney transplant recipients with COVID-19: A single center experience. *Transpl. Int.* **2020**, *33*, 1557–1561. [CrossRef]

59. Mohamed, I.H.; Chowdary, P.B.; Shetty, S.; Sammartino, C.; Sivaprakasam, R.; Lindsey, B.; Thuraisingham, R.; Yaqoob, M.M.; Khurram, M.A. Outcomes of renal transplant recipients with SARS-CoV-2 infection in the eye of the storm: A comparative study with waitlisted patients. *Transplantation* **2021**, *105*, 115–120. [CrossRef]

60. Molaei, H.; Khedmat, L.; Nemati, E.; Rostami, Z.; Saadat, S.H. Iranian kidney transplant recipients with COVID-19 infection: Clinical outcomes and cytomegalovirus coinfection. *Transpl. Infect. Dis.* **2021**, *23*, 13455. [CrossRef]

61. Monfared, A.; Dashti-Khavidaki, S.; Jafari, R.; Jafari, A.; Ramezanzade, E.; Lebadi, M.; Haghdar-Saheli, Y.; Aghajanzadeh, P.; Khosravi, M.; Movassaghi, A.; et al. Clinical characteristics and outcome of COVID-19 pneumonia in kidney transplant recipients in Razi hospital, Rasht, Iran. *Transpl. Infect. Dis.* **2020**, *22*, e13420. [CrossRef]

62. Montagud-Marrahi, E.; Cofan, F.; Torregrosa, J.; Cucchiari, D.; Ventura-Aguiar, P.; Revuelta, I.; Bodro, M.; Piñeiro, G.J.; Esforzado, N.; Ugalde, J.; et al. Preliminary data on outcomes of SARS-CoV-2 infection in a Spanish single center cohort of kidney recipients. *Arab. Archaeol. Epigr.* **2020**, *20*, 2958–2959. [CrossRef]

63. Nair, V.; Jandovitz, N.; Hirsch, J.S.; Nair, G.; Abate, M.; Bhaskaran, M.; Grodstein, E.; Berlinrut, I.; Hirschwerk, D.; Cohen, S.L.; et al. COVID-19 in kidney transplant recipients. *Arab. Archaeol. Epigr.* **2020**, *20*, 1819–1825. [CrossRef] [PubMed]

64. Oto, O.A.; Ozturk, S.; Turgutalp, K.; Arici, M.; Alpay, N.; Merhametsiz, O.; Sipahi, S.; Ogutmen, M.B.; Yelken, B.; Altiparmak, M.R.; et al. Predicting the outcome of COVID-19 infection in kidney transplant recipients. *BMC Nephrol.* **2021**, *22*, 100. [CrossRef] [PubMed]

65. Ozturk, S.; Turgutalp, K.; Arici, M.; Odabas, A.R.; Altiparmak, M.R.; Aydin, Z.; Cebeci, E.; Basturk, T.; Soypacaci, Z.; Sahin, G.; et al. Mortality analysis of COVID-19 infection in chronic kidney disease, haemodialysis and renal transplant patients compared with patients without kidney disease: A nationwide analysis from Turkey. *Nephrol. Dial. Transplant.* **2020**, *35*, 2083–2095. [CrossRef] [PubMed]

66. Phanish, M.; Ster, I.C.; Ghazanfar, A.; Cole, N.; Quan, V.; Hull, R.; Banerjee, D. Systematic review and meta-analysis of COVID-19 and kidney transplant recipients, the South West London kidney transplant network experience. *Kidney Int. Rep.* **2021**, *6*, 574–585. [CrossRef] [PubMed]

67. Pierrotti, L.C.; Junior, J.O.R.; Freire, M.P.; Machado, D.J.B.; Moreira, R.M.; Ventura, C.G.; Litvoc, M.N.; Nahas, W.C.; David-Neto, E. COVID-19 among kidney-transplant recipients requiring hospitalization: Preliminary data and outcomes from a single-center in Brazil. *Transpl. Int.* **2020**, *33*, 1837–1842. [CrossRef]

68. Rodriguez-Cubillo, B.; De La Higuera, M.A.M.; Lucena, R.; Franci, E.V.; Hurtado, M.; Romero, N.C.; Moreno, A.R.; Valencia, D.; Velo, M.; Fornie, I.S.; et al. Should cyclosporine be useful in renal transplant recipients affected by SARS-CoV-2? *Arab. Archaeol. Epigr.* **2020**, *20*, 3173–3181. [CrossRef]

69. De Sandes-Freitas, T.V.; Cristelli, M.P.; Neri, B.D.O.; Guedes, A.L.M.D.O.; Esmeraldo, R.D.M.; Garcia, V.D.; Prá, R.L.D.; Suassuna, J.H.R.; Rioja, S.D.S.; Zanocco, J.A.; et al. The unpredictable outcome of SARS-CoV-2 in kidney transplant recipients with HIV-infection. *Transplantation* **2021**, *105*, e9–e10. [CrossRef]

70. Santeusanio, A.D.; Menon, M.C.; Liu, C.; Bhansali, A.; Patel, N.; Mahir, F.; Rana, M.; Tedla, F.; Mahamid, A.; Fenig, Y.; et al. Influence of patient characteristics and immunosuppressant management on mortality in kidney transplant recipients hospitalized with coronavirus disease 2019 (COVID-19). *Clin. Transplant.* **2021**, *35*, e14221. [CrossRef]

71. Shrivastava, P.; Prashar, R.; Khoury, N.; Patel, A.; Yeddula, S.; Kitajima, T.; Nagai, S.; Samaniego, M. Acute kidney injury in a predominantly African American cohort of kidney transplant recipients with COVID-19 infection. *Transplantation* **2021**, *105*, 201–205. [CrossRef]

72. Tejada, C.D.J.; Zachariah, M.; Cruz, A.B.V.; Hussein, S.; Wipula, E.; Meeks, N.; Wolff, J.; Chandrasekar, P.H. Favorable outcome of COVID-19 among African American (AA) renal transplant recipients in Detroit. *Clin. Transplant.* **2021**, *35*, e14169. [CrossRef]

73. Trujillo, H.; Caravaca-Fontán, F.; Sevillano, Á.; Gutiérrez, E.; Fernández-Ruiz, M.; López-Medrano, F.; Hernández, A.; Aguado, J.M.; Praga, M.; Andrés, A. Tocilizumab use in kidney transplant patients with COVID-19. *Clin. Transplant.* **2020**, *34*. [CrossRef]

74. Villa, L.; Krüger, T.; Seikrit, C.; Mühlfeld, A.S.; Kunter, U.; Werner, C.; Kleines, M.; Schulze-Hagen, M.; Dreher, M.; Kersten, A.; et al. Time on previous renal replacement therapy is associated with worse outcomes of COVID-19 in a regional cohort of kidney transplant and dialysis patients. *Medicine* **2021**, *100*, e24893. [CrossRef]

75. Villanego, F.; Mazuecos, A.; Pérez-Flores, I.M.; Moreso, F.; Andrés, A.; Jiménez-Martín, C.; Molina, M.; Canal, C.; Sánchez-Cámara, L.A.; Zárraga, S.; et al. Predictors of severe COVID-19 in kidney transplant recipients in the different epidemic waves: Analysis of the Spanish Registry. *Arab. Archaeol. Epigr.* **2021**, *21*, 2573–2582. [CrossRef]

76. Willicombe, M.; Gleeson, S.; Clarke, C.; Dor, F.; Prendecki, M.; Lightstone, L.; Lucisano, G.; McAdoo, S.; Thomas, D.; on behalf of the ICHNT Renal COVID Group. Identification of Patient Characteristics Associated With SARS-CoV-2 Infection and Outcome in Kidney Transplant Patients Using Serological Screening. *Transplantation* **2021**, *105*, 151–157. [CrossRef]

77. Zhu, L.; Gong, N.; Liu, B.; Lu, X.; Chen, D.; Chen, S.; Shu, H.; Ma, K.; Xu, X.; Guo, Z.; et al. Coronavirus disease 2019 pneumonia in immunosuppressed renal transplant recipients: A summary of 10 confirmed cases in Wuhan, China. *Eur. Urol.* **2020**, *77*, 748–754. [CrossRef]

78. Meyerowitz-Katz, G.; Merone, L. A systematic review and meta-analysis of published research data on COVID-19 infection fatality rates. *Int. J. Infect. Dis.* **2020**, *101*, 138–148. [CrossRef]

79. Chen, Y.-T.; Shao, S.-C.; Hsu, C.-K.; Wu, I.-W.; Hung, M.-J.; Chen, Y.-C. Incidence of acute kidney injury in COVID-19 infection: A systematic review and meta-analysis. *Crit. Care* **2020**, *24*, 1–4. [CrossRef]

80. Levin, A.T.; Hanage, W.P.; Owusu-Boaitey, N.; Cochran, K.B.; Walsh, S.P.; Meyerowitz-Katz, G. Assessing the age specificity of infection fatality rates for COVID-19: Systematic review, meta-analysis, and public policy implications. *Eur. J. Epidemiol.* **2020**, *35*, 1123–1138. [CrossRef]

81. Gleeson, S.E.; Formica, R.N.; Marin, E.P. Outpatient management of the kidney transplant recipient during the SARS-CoV-2 virus pandemic. *Clin. J. Am. Soc. Nephrol.* **2020**, *15*, 892–895. [CrossRef]

82. Beran, A.; Zink, E.; Mhanna, M.; Abugharbyeh, A.; Do, J.H.; Duggan, J.; Assaly, R. Transmissibility and viral replication of SARS-COV-2 in immunocompromised patients. *J. Med. Virol.* **2021**, *93*, 4156–4160. [CrossRef]

83. Chan, S.; Pascoe, E.M.; Clayton, P.A.; McDonald, S.P.; Lim, W.H.; Sypek, M.P.; Palmer, S.C.; Isbel, N.M.; Francis, R.S.; Campbell, S.B.; et al. Infection-related mortality in recipients of a kidney transplant in Australia and New Zealand. *Clin. J. Am. Soc. Nephrol.* **2019**, *14*, 1484–1492. [CrossRef] [PubMed]

84. Grasselli, G.; Greco, M.; Zanella, A.; Albano, G.; Antonelli, M.; Bellani, G.; Bonanomi, E.; Cabrini, L.; Carlesso, E.; Castelli, G.; et al. Risk factors associated with mortality among patients with COVID-19 in intensive care units in Lombardy, Italy. *JAMA Intern. Med.* **2020**, *180*, 1345. [CrossRef] [PubMed]

85. Kim, L.; Garg, S.; O'Halloran, A.; Whitaker, M.; Pham, H.; Anderson, E.J.; Armistead, I.; Bennett, N.M.; Billing, L.; Como-Sabetti, K.; et al. Risk factors for intensive care unit admission and in-hospital mortality among hospitalized adults identified through the US coronavirus disease 2019 (COVID-19)-associated hospitalization surveillance network (COVID-NET). *Clin. Infect. Dis.* **2021**, *72*, e206–e214. [CrossRef] [PubMed]

86. Atkins, J.L.; Masoli, J.A.H.; Delgado, J.; Pilling, L.C.; Kuo, C.-L.; Kuchel, G.A.; Melzer, D. Preexisting comorbidities predicting COVID-19 and mortality in the UK biobank community cohort. *J. Gerontol. Ser. A Biol. Sci. Med. Sci.* **2020**, *75*, 2224–2230. [CrossRef]

87. Cardinal, H.; Dieudé, M.; Hébert, M.-J. Endothelial dysfunction in kidney transplantation. *Front. Immunol.* **2018**, *9*, 1130. [CrossRef]

88. Chen, Y.-C.; Fang, J.-T.; Yang, C.-W. Endothelial dysfunction for acute kidney injury in coronavirus disease 2019: How concerned should we be? *Nephron* **2021**, *145*, 513–517. [CrossRef]

89. A Kellum, J.; van Till, J.W.O.; Mulligan, G. Targeting acute kidney injury in COVID-19. *Nephrol. Dial. Transplant.* **2020**, *35*, 1652–1662. [CrossRef]

90. Chang, R.; Elhusseiny, K.M.; Yeh, Y.-C.; Sun, W.-Z. COVID-19 ICU and mechanical ventilation patient characteristics and outcomes—A systematic review and meta-analysis. *PLoS ONE* **2021**, *16*, e0246318. [CrossRef]
91. Thakur, B.; Dubey, P.; Benitez, J.; Torres, J.P.; Reddy, S.; Shokar, N.; Aung, K.; Mukherjee, D.; Dwivedi, A.K. A systematic review and meta-analysis of geographic differences in comorbidities and associated severity and mortality among individuals with COVID-19. *Sci. Rep.* **2021**, *11*, 1–13. [CrossRef]
92. Basuki, W.; Pramudya, D.; Adiwinoto, R.D. Cyclosporine a improves outcome of kidney transplant recipients with coronav irus disease 2019: A meta-analysis. *New Armen. Med. J.* **2020**, *14*, 100–106.

Journal of
Clinical Medicine

Article

The Effects of Income Level on Susceptibility to COVID-19 and COVID-19 Morbidity/Mortality: A Nationwide Cohort Study in South Korea

So Young Kim [1], Dae Myoung Yoo [2], Chanyang Min [2,3] and Hyo Geun Choi [2,4,*]

1   Department of Otorhinolaryngology-Head & Neck Surgery, CHA Bundang Medical Center, CHA University, Seongnam 13496, Korea; sossi81@hanmail.net
2   Hallym Data Science Laboratory, Hallym University College of Medicine, Anyang 14068, Korea; ydm1285@naver.com (D.M.Y.); joicemin@naver.com (C.M.)
3   Graduate School of Public Health, Seoul National University, Seoul 08826, Korea
4   Department of Otorhinolaryngology-Head & Neck Surgery, Hallym University College of Medicine, Anyang 14068, Korea
*   Correspondence: pupen@naver.com

Abstract: This study aimed to investigate the association of income level with susceptibility to coronavirus disease 2019 (COVID-19) and COVID-19 morbidity and mortality. Using the Korean National Health Insurance COVID-19 Database cohort, medical claim data from 2015 through 2020 were collected. A total of 7943 patients who were diagnosed with COVID-19 from 1 January 2020 to 4 June 2020 were included. A total of 118,914 participants had negative COVID-19 PCR tests. Income levels were classified by 20th percentiles based on 2019 Korean National Health Insurance premiums. The 20th percentile income levels were categorized into three groups (low, middle, and high). The relationship of income level with susceptibility to COVID-19 and COVID-19 morbidity and mortality was analyzed using logistic regression analysis. A high income level was related to lower odds of COVID-19 infection (adjusted odds ratio (aOR) = 0.79, 95% confidence interval (CI) = 0.75–0.83, $p < 0.001$). The negative association between income level and COVID-19 infection was maintained in all subgroups. Patients with low income levels were susceptible to COVID-19 infection; however, there was no relation of COVID-19 morbidity and mortality with income level in the Korean population.

Keywords: healthcare disparities; morbidity; mortality; COVID-19; case-control studies; cohort studies

Citation: Kim, S.Y.; Yoo, D.M.; Min, C.; Choi, H.G. The Effects of Income Level on Susceptibility to COVID-19 and COVID-19 Morbidity/Mortality: A Nationwide Cohort Study in South Korea. *J. Clin. Med.* **2021**, *10*, 4733. https://doi.org/10.3390/jcm10204733

Academic Editor: Yoav Yinon

Received: 26 August 2021
Accepted: 13 October 2021
Published: 15 October 2021

Publisher's Note: MDPI stays neutral with regard to jurisdictional claims in published maps and institutional affiliations.

1. Introduction

The coronavirus disease 2019 (COVID-19) pandemic has greatly affected many aspects of life for people around the world. Health resources were redistributed and focused on coping with severe acute respiratory syndrome coronavirus 2 (SARS-CoV-2) infection [1]. The social lockdown and restricted economic activities exposed many people to unemployment and bankruptcy. People in occupations requiring physical labor or face-to-face services, such as food service employees, supermarket or warehouse workers, telemarketers, and drivers of public transportation, selectively encountered unprecedented job loss. Conversely, individuals with contact-free jobs, such as programmers, executive officers, and capitalists, could continue their jobs from home and were less influenced by the COVID-19 epidemic. As a result, economic inequalities have been accentuated during the COVID-19 epidemic [2,3]. Increased economic inequalities are directly connected to health inequalities [4]. The disparity of income is closely related to many factors associated with socioeconomic risks, which may contribute to susceptibility to infection and COVID-19 mortality [4].

Low economic status has been highlighted as a factor affecting vulnerability to COVID-19 infection [5–7]. Patients with a low economic status are more likely to reside in unhealthy

73

environments with poor hygiene. Workplaces and living spaces are unfavorable for maintaining social distancing to combat viral transmission. In addition, a diminished food industry causes shortages of food and food insecurity, which increases the risk of COVID-19 infection in people living in poverty [8]. The reproductive ratios of COVID-19 were as high as 1.29 (95% confidence intervals (CI) = 1.15–1.46) in groups with unfavorable socioeconomic status, higher than the median of 0.96 (interquartile range = 0.72–1.34) [9]. Communities with lower incomes, less insurance coverage, and more unemployment were associated with higher rates of COVID-19 in an ecological study [7]. Moreover, a low economic status may impede the early diagnosis and treatment of COVID-19, which increases the severity of disease [10]. In an ecological study, both the incidence and mortality of COVID-19 were correlated with the Gini coefficient (rho = +0.6906, p < 0.001 for the incidence of COVID-19 and rho = +0.6564, p < 0.001 for the mortality of COVID-19) [6]. To assess the relation of COVID-19 infection with economic status, other socioeconomic factors, including ethnicity, region of residence, and health insurance system, should be included in the analyses. In Korea, the diagnosis and treatment costs of COVID-19 have been completely covered by the Korean government, regardless of patients' economic status. Thus, the Korean cohort excludes the influence of accessibility and availability of diagnosis and treatment of COVID-19 across economic levels.

We hypothesized that income level could have an impact on susceptibility to COVID-19 and on the morbidity and mortality of COVID-19. To minimize potential confounding effects, the analysis was adjusted for the covariates age, sex, and comorbidity.

2. Materials and Methods

2.1. Ethics

The Ethics Committee of Hallym University (2020-07-022) permitted this study. Written informed consent was waived by the Institutional Review Board. The participants' information was anonymized. All analyses adhered to the guidelines and regulations of the Ethics Committee of Hallym University.

2.2. Study Population and Participant Selection

We used the Korean National Health Insurance COVID-19 Database (NHID-COVID DB) medical claim code data from 2015 to 2020. The NHID-COVID DB provided data for individuals who underwent SARS-CoV-2 testing, using real-time reverse transcriptase PCR assay of nasal or pharyngeal swabs, in accordance with the WHO guidelines. Control participants from the Korean National Health Insurance Database were matched by age and sex.

Confirmed COVID-19 patients were included from 1 January 2020 to 4 June 2020; all of them finished treatment or died by 4 June 2020 (n = 8070). Fifteen times more control participants matched by age and sex were extracted (n = 121,050). Among them, we excluded participants with a lack of income records (n = 127 for COVID-19 patients, n = 2136 for control participants). Consequently, 7943 COVID-19 participants and 118,914 control participants were selected. Then, COVID-19 patients were analyzed for mild (n = 7385) and severe (n = 558) morbidity. They were also analyzed for death (n = 233) and survival (n = 7710) (Figure 1).

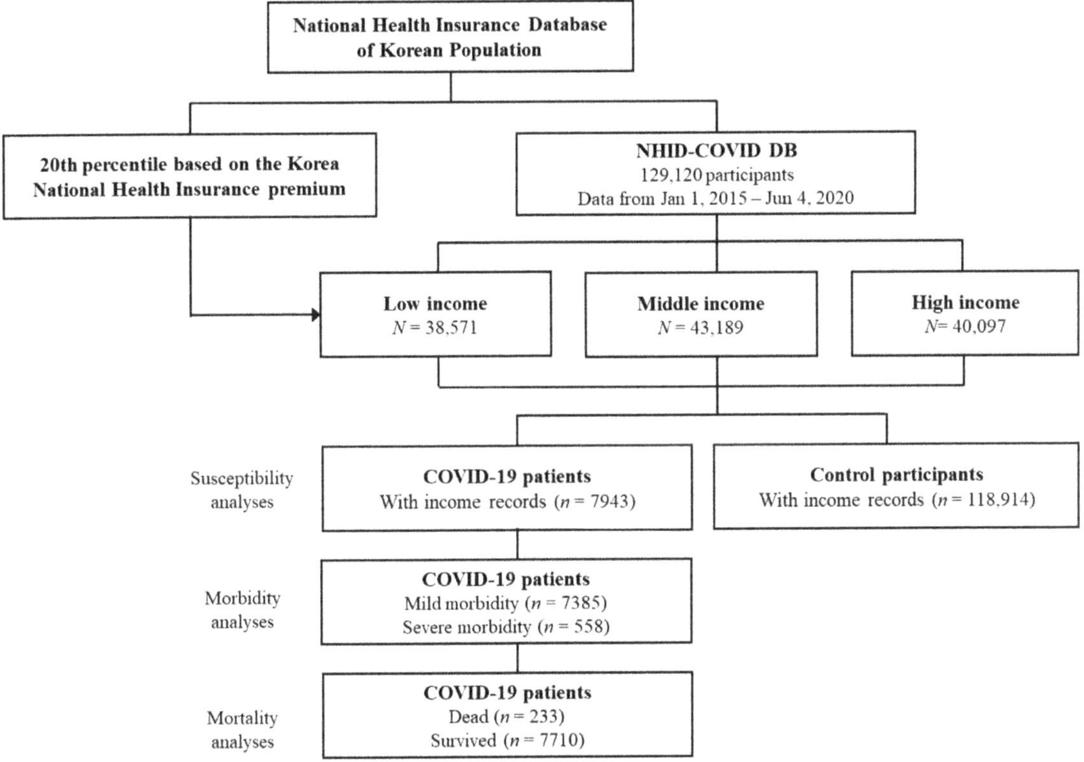

Figure 1. A schematic illustration of the participant selection process that was used in the present study. Of a total of 129,120 participants, 7943 COVID-19 patients and 118,914 control participants were selected.

2.3. Exposure (Income Level)

Income level was divided into 20th percentiles based on 2019 Korea National Health Insurance premiums, ranging from 1 (the lowest 5%) to 20 (the highest 5%), for the entire Korean population with health insurance (Supplementary Table S1) [11]. In addition, medical-aid beneficiaries were added to the lowest income level, which was estimated to be approximately 3.0% of the total Korean population (class 0) [11,12]. We categorized income level into 3 groups (low (income level 0 to 6), middle (income level 7 to 14), and high (income level 15 to 20)).

2.4. Outcome (COVID-19 Infection)

Laboratory confirmation of SARS-CoV-2 infection, using a real-time reverse transcriptase PCR assay, was defined as the primary outcome.

2.5. Secondary Outcome (Morbidity and Mortality)

The secondary outcomes were morbidity and mortality in COVID-19 patients. Morbidity was defined as mild or severe. Severe morbidity was indicated by admission to the intensive care unit (ICU), invasive ventilation, extracorporeal membrane oxygenation (ECMO), or death.

2.6. Covariates

Age groups were divided into 10-year intervals: 0–9, 10–19, 20–29 and so on, with the oldest group being 80+ years old (total of 9 age groups).

The Charlson comorbidity index (CCI) has been widely used to measure disease burden using 17 comorbidities: myocardial infarction, congestive heart failure, peripheral vascular disease, hemiplegia or paraplegia, dementia, chronic pulmonary disease, rheumatologic disease, peptic ulcer disease, diabetes without chronic complications, diabetes with chronic complications, renal disease, any malignancy, including leukemia and lymphoma, metastatic solid tumor, mild liver disease, moderate or severe liver disease, and HIV/AIDS [13]. The presence of each comorbidity was counted with a weighted value and summed as a CCI score. It is a continuous variable (0 (no comorbidities) through 29 (multiple comorbidities)) [13]. In addition, hypertension (ICD-10 codes: I10 and I15) was assigned if participants were treated $\geq$2 times, as it was not included in the CCI.

2.7. Statistical Analyses

The general characteristics of all participants were compared among income groups using the chi-squared test.

To estimate the susceptibility to COVID-19, of COVID-19 patients compared to control participants, odds ratios (ORs) with 95% confidence intervals (CIs) of income were calculated using crude (simple model) and adjusted (for age, sex, CCI score, and hypertension) logistic regression models. To estimate morbidity/mortality in COVID-19 patients by income, logistic regression was used. For subgroup analyses, we divided participants by age (<50 years old and $\geq$50 years old), sex, CCI score (0 score, 1 score, and $\geq$2 score), and hypertension history.

For the statistical analyses, SAS version 9.4 (SAS Institute Inc., Cary, NC, USA) was used. We performed two-tailed analyses, and significance was defined as p values less than 0.05.

3. Results

The prevalence of COVID-19 was different among income groups (p < 0.001, Table 1). Totals of 7.4% (2836/38,571), 5.8% (2489/43,189), and 5.8% (2618/40,097) of the low-, middle-, and high-income groups had histories of COVID-19. The morbidity of COVID-19 was 6.5% (185/2836), 6.5% (161/2489), and 8.1% (212/2618) for the low-, middle-, and high-income groups, respectively (p = 0.03). The mortality of COVID-19 was 0.23% (86/2836), 0.14% (62/2489), and 0.19% (85/2618) for the low-, middle-, and high-income groups, respectively (p = 0.03). The distributions of age, sex, CCI score, and history of hypertension were different among income groups (all p < 0.001).

Income level was inversely related to susceptibility to COVID-19 (Table 2). Compared to the low-income group, the middle- and high-income groups demonstrated lower odds of COVID-19 infection (adjusted OR (aOR) = 0.78, 95% CI = 0.74–0.83, p < 0.001 for the middle-income group and aOR = 0.79, 95% CI = 0.75–0.83, p < 0.001 for the high-income group). According to the analysis of 20th percentile income levels, ranging from 1 (the lowest 5%) to 20 (the highest 5%), a high income level was associated with 0.98 times lower odds of COVID-19 infection (95% CI = 0.98–0.99, p < 0.001). Additional analyses according to age, sex, CCI score, and history of hypertension showed a consistent association of COVID-19 infection with lower income (Supplementary Table S2).

Table 1. General characteristics of the participants.

Characteristics	Total Participants			
	Low-Income Group (*n*, %)	Middle-Income Group (*n*, %)	High-Income Group (*n*, %)	*p*-Value
Total number	38,571 (100.0)	43,189 (100.0)	45,097 (100.0)	
Age (years old)				<0.001 *
0–9	238 (0.6)	471 (1.1)	533 (1.2)	
10–19	996 (2.6)	1207 (2.8)	2107 (4.7)	
20–29	10,810 (28.0)	11,946 (27.7)	9560 (21.2)	
30–39	3431 (8.9)	5907 (13.7)	3580 (7.9)	
40–49	4792 (12.4)	5338 (12.4)	6164 (13.7)	
50–59	7958 (20.6)	8366 (19.4)	8476 (18.8)	
60–69	5977 (15.5)	6167 (14.3)	6732 (14.9)	
70–79	2469 (6.4)	2375 (5.5)	4865 (10.8)	
80+	1900 (4.9)	1412 (3.3)	3080 (6.8)	
Sex				<0.001 *
Male	13,716 (35.6)	17,847 (41.3)	19,216 (42.6)	
Female	24,855 (64.4)	25,342 (58.7)	25,881 (57.4)	
CCI score				<0.001 *
0	34,603 (89.7)	39,946 (92.5)	40,654 (90.2)	
1	2028 (5.3)	1743 (4.0)	2314 (5.1)	
≥2	1940 (5.0)	1500 (3.5)	2129 (4.7)	
Hypertension	7888 (20.5)	7552 (17.5)	10,257 (22.7)	<0.001 *
COVID-19	2836 (7.4)	2489 (5.8)	2618 (5.8)	<0.001 *
Prognosis of COVID-19				
Morbidity	185 (6.5)	161 (6.5)	212 (8.1)	0.032 *
Mortality	86 (0.23)	62 (0.14)	85 (0.19)	0.029 *

Abbreviations: CCI, Charlson comorbidity index; COVID-19, coronavirus disease 2019. * Chi-squared test. Significance at *p* < 0.05.

Table 2. Crude and adjusted odds ratios of the association of income with COVID-19 infection in the total participants.

Characteristics	COVID-19	Control	ORs (95% Confidence Interval) for COVID-19			
	(Exposure/Total, %)	(Exposure/Total, %)	Crude	*p*-Value	Adjusted [†]	*p*-Value
Income group						
Low	2836/7943 (35.7%)	35,735/118,914 (30.1%)	1		1	
Middle	2489/7943 (31.3%)	40,700/118,914 (34.2%)	0.77 (0.73–0.82)	<0.001 *	0.78 (0.74–0.83)	<0.001 *
High	2618/7943 (33.0%)	42,479/118,914 (35.7%)	0.78 (0.74–0.82)	<0.001 *	0.79 (0.75–0.83)	<0.001 *
Income level (mean, SD)	10.00 (6.76)	10.75 (6.39)	0.98 (0.98–0.99)	<0.001 *	0.98 (0.98–0.99)	<0.001 *

* Logistic regression model, significance at *p* < 0.05. [†] Adjusted model for age, sex, CCI score and hypertension.

The morbidity of COVID-19 was not associated with income level in the adjusted models (Table 3). The high-income group showed 1.26 times higher odds of COVID-19 morbidity in the crude model (95% CI = 1.03–1.55, *p* = 0.03); however, there was no significant association of COVID-19 morbidity with income level when adjusted for age, sex, CCI score, and hypertension. Among the age, sex, CCI score, and history of hypertension subgroups, males with no past medical history (CCI score = 0), and an absence of hypertension, had higher odds of COVID-19 morbidity in the higher income groups (Supplement Table S3). The middle-income level demonstrated 1.49 times higher odds of COVID-19 morbidity than the low-income level in the male group (95% CI = 1.06–2.07,

$p = 0.03$). The group with no past medical history and the hypertension-free group showed 1.03 times (95% CI = 1.01–1.05, $p = 0.004$) and 1.02 times (95% CI = 1.00–1.04, $p = 0.03$) higher odds of COVID-19 morbidity with higher income levels, respectively.

Table 3. Crude and adjusted odds ratios of the association of income with morbidity in COVID-19 participants.

Characteristics	Severe Participants	Mild Participants	ORs (95% Confidence Interval) for Morbidity			
	(Exposure/Total, %)	(Exposure/Total, %)	Crude	*p*-Value	Adjusted [†]	*p*-Value
	Income group					
Low	185/558 (33.2%)	2651/7385 (35.9%)	1		1	
Middle	161/558 (28.9%)	2328/7385 (31.5%)	0.99 (0.80–1.23)	0.936	1.21 (0.96–1.53)	0.108
High	212/558 (38.0%)	2406/7385 (32.6%)	1.26 (1.03–1.55)	0.026 *	1.17 (0.94–1.46)	0.172
Income level (mean, SD)	10.64 (7.21)	9.95 (6.73)	1.02 (1.00–1.03)	0.020 *	1.01 (1.00–1.03)	0.056

* Logistic regression model, significance at $p < 0.05$. [†] Adjusted model for age, sex, CCI score and hypertension.

COVID-19 mortality was not associated with income level (Table 4). Neither the middle- nor high-income groups showed increased odds of mortality due to COVID-19 (aOR = 1.09, 95% CI = 0.75–1.58, $p = 0.65$ for the middle-income group and aOR = 0.76, 95% CI = 0.54–1.08, $p = 0.19$ for the high-income group). None of the 20th percentile income levels were related to mortality due to COVID-19 (aOR = 0.99, 95% CI = 0.97–1.01, $p = 0.15$). None of the age, sex, CCI score, or history of hypertension subgroups showed an association between COVID-19 mortality and income level, except for the group with a CCI score = 1 (Supplementary Table S4). In the CCI score = 1 group, the high-income group had 0.43 times lower odds of mortality due to COVID-19 (95% CI = 0.22–0.83, $p = 0.01$).

Table 4. Crude and adjusted odds ratios of the association of income with mortality in COVID-19 participants.

Characteristics	Dead Participants	Survived Participants	ORs (95% Confidence Interval) for Mortality			
	(Exposure/Total, %)	(Exposure/Total, %)	Crude	*p*-Value	Adjusted [†]	*p*-Value
	Income group					
Low	86/233 (36.9%)	2750/7710 (35.7%)	1		1	
Middle	62/233 (26.6%)	2427/7710 (31.5%)	0.82 (0.59–1.14)	0.231	1.09 (0.75–1.58)	0.654
High	85/233 (36.5%)	2533/7710 (32.9%)	1.07 (0.79–1.46)	0.650	0.76 (0.54–1.08)	0.123
Income level (mean, SD)	10.07 (7.57)	10.00 (6.74)	1.00 (0.98–1.02)	0.876	0.99 (0.97–1.01)	0.148

[†] Adjusted model for age, sex, CCI score and hypertension.

4. Discussion

4.1. Principal Results

A lower income level was associated with a higher susceptibility to COVID-19 infection; however, COVID-19 morbidity and mortality were not related to income level in the overall population. The mortality of COVID-19 was lower in the high-income group in the CCI score = 1 subgroup. However, the morbidity of COVID-19 was higher at high income levels in the male sex, CCI score = 0, and hypertension-free subgroups. The present results indicated an increased susceptibility to COVID-19 infection in lower-income-level participants; therefore, a correlation mostly likely exists between economic inequality and COVID-19 susceptibility. The present study improved upon previous studies by analyzing susceptibility to COVID-19 and COVID-19 morbidity and mortality in the same national cohort. This study examined the impact of economic level on susceptibility to COVID-19 and COVID-19 morbidity and mortality in the absence of disparities in the availability of medical resources.

4.2. Comparison with Prior Work

A number of previous studies suggested a higher susceptibility to COVID-19 infection in lower economic groups [5–7,9,14]. A retrospective study in a European urban area showed increased incidences of COVID-19 in low-income groups (risk ratio (RR) = 1.67, 95% CI = 1.41–1.96 for men and RR = 1.71, 95% CI = 1.44–1.99 for women) [14]. A low income level could influence susceptibility to COVID-19 via an elevated risk of viral exposure and an immune system that is impaired during the neutralization of a viral infection. An increased possibility of viral exposure could be linked to a higher risk of COVID-19 infection in the low-income population. Adverse living and working environments may increase the risk of COVID-19 infection in low-income populations. Poverty and one's physical environment, such as a homeless status and/or exposure to smoking, are social determinants of health and have an impact on COVID-19 outcomes [15]. Crowded living conditions, poor hygiene, less access to healthcare, and quarantining can increase the risk of viral infection in homeless populations [16]. The low-income group exhibited less social distancing during the COVID-19 pandemic [17]. As much as approximately 36.0% (147/408) of the homeless population in Boston tested positive for SARS-CoV-2, using PCR testing [18]. The group with a low socioeconomic status demonstrated a strong association of COVID-19 infection with current smoking (aOR = 3.53, 95% CI = 1.22–2.62) [19].

In the present study, the income level was classified based on the health insurance premium, which reflected the income quintile. All Koreans must be registered with the national health insurance system; therefore, the classified income levels were precise. Korea was ranked as the country with the 12th highest gross domestic product (GDP) worldwide in 2017 [20]. Compared to other countries with similar GDP levels, such as Italy, Australia, and Spain, Korea showed a lower rate of contraction of SARS-CoV-2 and a lower mortality rate for COVID-19. A number of features, including a strong central autonomous agency that used research for agile and responsive policymaking, public trust in government measures, strong public–private sector collaboration, and surveillance and response built on integrated information management systems, could contribute to the lower infection rate and mortality rate of COVID-19 in Korea [21]. In addition, the Korean government covered all medical costs for COVID-19, enabling all participants to be examined and treated without discrimination. Additional factors contributing to socioeconomic deprivation could affect susceptibility to COVID-19, such as occupation, educational level, housing status, and food security, which were not available in the present cohort [22]. Another Korean epidemiological study suggested increased susceptibility to COVID-19 in participants with less healthcare access, less education, more risky health behaviors, crowding, specific comorbidities, difficulty social distancing, and population mobility [23].

Decreased immune system ability to combat SARS-CoV-2 infection could increase susceptibility to COVID-19 in low-income populations. Pre-existing health inequalities could add to the risk of COVID-19 infection in low-income populations. Low socioeconomic status was associated with a higher rate of chronic diseases, which made individuals with that status more vulnerable to COVID-19 [24]. Comorbidities, including diabetes and kidney diseases, have been associated with higher COVID-19 morbidity [25,26]. The overall comorbidity burdens were estimated to be approximately 1.3 times higher for hospitalization for COVID-19 in white patients (95% CI = 1.11–1.53, p = 0.001) [25]. In addition, an increased stress level may diminish immune functioning in the low-income group [5]. A weakened immune system could increase invasion by and replication of SARS-CoV-2 in this population. Low socioeconomic status was associated with perceived stress and health-risk behavior in a cross-sectional study (aOR = 2.90, 95% CI = 2.53–3.33 for perceived stress) [27], and the COVID-19 epidemic is likely to have a higher impact on the economic status of low-income groups because it may impose higher stress on these groups. In fact, during the COVID-19 pandemic, the lower-income group developed severe psychological distress more often than the higher-income group in a longitudinal study (aOR = 3.00, 95% CI = 1.01–9.58) [28]. Acute stress and chronic stress tended to suppress cellular and humoral immunity in a meta-analysis [29]. Thus, high stress in the low-income group could increase susceptibility to COVID-19.

The morbidity and mortality of COVID-19 did not show an association with income level in this study. In contrast, several retrospective studies reported a higher risk of severe illness from COVID-19 in low-income populations [30,31]. In a U.S. study, the low-income group had a higher risk of severe illness from COVID-19 than the higher-income groups (prevalence ratios = 1.63, 95% CI = 1.59–1.67) [30]. Moreover, the initial severity of COVID-19 was higher in patients residing in a poor district of Paris (aOR = 1.099, 95% CI = 1.038–1.178) [31]. Full coverage of COVID-19 treatment costs may have minimized the cases of undertreatment in our cohort. In Korea, the medical costs related to the diagnosis and treatment of COVID-19 have been covered by the Korean government. The indemnity of insurance coverage was suggested to improve the opportunity for regular healthcare compared to the uninsured population [32]. In a propensity-score-matched case-control study, uninsured adults showed higher mortality than insured adults (adjusted hazard ratio = 1.43, 95% CI = 1.10–1.85, *p* = 0.01) [33]. Thus, patient income levels are unlikely to affect the procedures involved in COVID-19 therapy. Moreover, the relatively small number of COVID-19 cases with morbidity and mortality may attenuate the statistical power to delineate the association of income level with morbidity and mortality.

4.3. Limitations

The present study used a nationwide, representative cohort. Our cohort comprised a single ethnicity (Korean); therefore, the possible impacts of ethnic disparities on outcomes were minimized [34]. In addition, the bias from undetected or undertreated COVID-19 cases was likely minimized in our cohort because the Korean government diagnosed and treated COVID-19 without any charge. Healthcare resources were never in short supply in Korea, and the infection rate of SARS-CoV-2 was controlled at less than 2000 persons per day. Thus, the diagnosis and treatment of COVID-19 were not influenced by individual economic status in this study. However, a few limitations should be considered when applying the present results. Although adjustments were made for age, sex, and comorbidities, confounders for COVID-19 infection remained, such as occupation and region of residence. Information on occupation and region of residence was not available in the NHID-COVID DB to guarantee the participants' anonymity. These remaining confounders could have influenced the positive association of COVID-19 morbidity with high income levels in some subgroups in this study. This study included patients with COVID-19 from 1 January 2020 to 4 June 2020. This period was in the early COVID-19 pandemic period; therefore, the long-term effects of income level on COVID-19 infection need to be evaluated in further studies.

5. Conclusions

COVID-19 infection was higher in participants with lower income levels in the Korean population; however, the mortality of COVID-19 was not different according to income level in Korea. Public and government management of COVID-19 may impact the association of COVID-19 with income level. Health inequalities can be aggravated by a high rate of COVID-19 infection in deprived populations; therefore, active and prompt measures are essential.

Supplementary Materials: The following are available online at https://www.mdpi.com/article/10.3390/jcm10204733/s1, Supplementary Table S1: Distribution of national health insurance contributions in South Korea by income quintile. Supplementary Table S2: Subgroup analyses of crude and adjusted odds ratios of the association of income with COVID-19 infection in total participants by covariates. Supplementary Table S3: Subgroup analyses of crude and adjusted odds ratios of the association of income with morbidity in COVID-19 participants by covariates. Supplementary Table S4: Subgroup analyses of crude and adjusted odds ratios of the association of income with mortality in COVID-19 participants by covariates.

Author Contributions: H.G.C. designed the study; D.M.Y., C.M. and H.G.C. analyzed the data; S.Y.K. and H.G.C. drafted and revised the paper; and H.G.C. drew the figures. All authors have read and agreed to the published version of the manuscript.

Funding: This work was supported in part by research grants (NRF-2018-R1D1A1A02085328 and 2021-R1C1C100498611) from the National Research Foundation (NRF) of Korea. The APC was funded by NRF-2021-R1C1C100498611.

Institutional Review Board Statement: The Ethics Committee of Hallym University (2020-07-022) permitted this study following the guidelines and regulations.

Informed Consent Statement: Written informed consent was waived by the Institutional Review Board.

Data Availability Statement: Releasing the data by the researcher is not legally permitted. All data are available from the Korea Centers for Disease Control and Prevention database. The Korea Centers for Disease Control and Prevention database allows data access, at a particular cost, for any researcher who promises to follow the research ethic guidelines. The data of this article can be downloaded from the website after agreeing to follow the research ethic guidelines.

Conflicts of Interest: The authors declare no conflict of interest.

References

1. Wang, Z.; Tang, K. Combating COVID-19: Health equity matters. *Nat. Med.* **2020**, *26*, 458. [CrossRef]
2. Okonkwo, N.E.; Aguwa, U.T.; Jang, M.; Barré, I.A.; Page, K.R.; Sullivan, P.S.; Beyrer, C.; Baral, S. COVID-19 and the US response: Accelerating health inequities. *BMJ Evid. Based Med.* **2020**, *26*, 176–179. [CrossRef]
3. Dorn, A.V.; Cooney, R.E.; Sabin, M.L. COVID-19 exacerbating inequalities in the US. *Lancet* **2020**, *395*, 1243–1244. [CrossRef]
4. Wildman, J. COVID-19 and income inequality in OECD countries. *Eur. J. Health Econ.* **2021**, *22*, 455–462. [CrossRef]
5. Patel, J.A.; Nielsen, F.B.H.; Badiani, A.A.; Assi, S.; Unadkat, V.A.; Patel, B.; Ravindrane, R.; Wardle, H. Poverty, inequality and COVID-19: The forgotten vulnerable. *Public Health* **2020**, *183*, 110–111. [CrossRef]
6. Demenech, L.M.; Dumith, S.C.; Vieira, M.; Neiva-Silva, L. Income inequality and risk of infection and death by COVID-19 in Brazil. *Rev. Bras. Epidemiol.* **2020**, *23*, e200095. [CrossRef]
7. Hawkins, D. Social Determinants of COVID-19 in Massachusetts, United States: An Ecological Study. *J. Prev. Med. Public Health* **2020**, *53*, 220–227. [CrossRef]
8. Pereira, M.; Oliveira, A.M. Poverty and food insecurity may increase as the threat of COVID-19 spreads. *Public Health Nutr.* **2020**, *23*, 3236–3240. [CrossRef]
9. Breitling, L.P. Global epidemiology and socio-economic development correlates of the reproductive ratio of COVID-19. *Int. Health* **2021**, ihab006. [CrossRef]
10. Szczepura, A. Access to health care for ethnic minority populations. *Postgrad. Med. J.* **2005**, *81*, 141–147. [CrossRef]
11. National Health Insurance Statistical Yearbook. 2019. Available online: http://www.hira.or.kr/bbsDummy.do?pgmid=HIRAA020045020000 (accessed on 2 June 2021).
12. Deug-Soo Kim, J.A. Characteristics in Atmospheric Chemistry between NO, NO_2 and O_3 at an Urban Site during MAPS (Megacity Air Pollution Study)-Seoul, Korea. *J. Korean Soc. Atmos. Environ.* **2016**, *32*, 422–434. [CrossRef]
13. Quan, H.; Li, B.; Couris, C.M.; Fushimi, K.; Graham, P.; Hider, P.; Januel, J.M.; Sundararajan, V. Updating and validating the Charlson comorbidity index and score for risk adjustment in hospital discharge abstracts using data from 6 countries. *Am. J. Epidemiol.* **2011**, *173*, 676–682. [CrossRef]
14. Mari-Dell'Olmo, M.; Gotsens, M.; Pasarin, M.I.; Rodriguez-Sanz, M.; Artazcoz, L.; Garcia de Olalla, P.; Rius, C.; Borrell, C. Socioeconomic Inequalities in COVID-19 in a European Urban Area: Two Waves, Two Patterns. *Int. J. Environ. Res. Public Health* **2021**, *18*, 1256. [CrossRef]
15. Abrams, E.M.; Szefler, S.J. COVID-19 and the impact of social determinants of health. *Lancet Respir. Med.* **2020**, *8*, 659–661. [CrossRef]
16. Tsai, J.; Wilson, M. COVID-19: A potential public health problem for homeless populations. *Lancet Public Health* **2020**, *5*, e186–e187. [CrossRef]
17. Weill, J.A.; Stigler, M.; Deschenes, O.; Springborn, M.R. Social distancing responses to COVID-19 emergency declarations strongly differentiated by income. *Proc. Natl. Acad. Sci. USA* **2020**, *117*, 19658–19660. [CrossRef]
18. Baggett, T.P.; Keyes, H.; Sporn, N.; Gaeta, J.M. Prevalence of SARS-CoV-2 Infection in Residents of a Large Homeless Shelter in Boston. *JAMA* **2020**, *323*, 2191–2192. [CrossRef] [PubMed]
19. Jackson, S.E.; Brown, J.; Shahab, L.; Steptoe, A.; Fancourt, D. COVID-19, smoking and inequalities: A study of 53,002 adults in the UK. *Tob. Control* **2020**. [CrossRef]

20. GDP by Country. Available online: https://www.worldometers.info/gdp/gdp-by-country/ (accessed on 2 June 2021).
21. *Assessment of COVID-19 Response in the Republic of Korea*; Asian Development Bank: Manila, Philippines, 2021.
22. Upshaw, T.L.; Brown, C.; Smith, R.; Perri, M.; Ziegler, C.; Pinto, A.D. Social determinants of COVID-19 incidence and outcomes: A rapid review. *PLoS ONE* **2021**, *16*, e0248336. [CrossRef] [PubMed]
23. Weinstein, B.; da Silva, A.R.; Kouzoukas, D.E.; Bose, T.; Kim, G.J.; Correa, P.A.; Pondugula, S.; Lee, Y.; Kim, J.; Carpenter, D.O. Precision Mapping of COVID-19 Vulnerable Locales by Epidemiological and Socioeconomic Risk Factors, Developed Using South Korean Data. *Int. J. Environ. Res. Public Health* **2021**, *18*, 604. [CrossRef] [PubMed]
24. Gray, D.M., II; Anyane-Yeboa, A.; Balzora, S.; Issaka, R.B.; May, F.P. COVID-19 and the other pandemic: Populations made vulnerable by systemic inequity. *Nat. Rev. Gastroenterol. Hepatol.* **2020**, *17*, 520–522. [CrossRef] [PubMed]
25. Gu, T.; Mack, J.A.; Salvatore, M.; Prabhu Sankar, S.; Valley, T.S.; Singh, K.; Nallamothu, B.K.; Kheterpal, S.; Lisabeth, L.; Fritsche, L.G.; et al. Characteristics Associated with Racial/Ethnic Disparities in COVID-19 Outcomes in an Academic Health Care System. *JAMA Netw. Open* **2020**, *3*, e2025197. [CrossRef]
26. Park, J.H.; Jang, W.; Kim, S.W.; Lee, J.; Lim, Y.S.; Cho, C.G.; Park, S.W.; Kim, B.H. The Clinical Manifestations and Chest Computed Tomography Findings of Coronavirus Disease 2019 (COVID-19) Patients in China: A Proportion Meta-Analysis. *Clin. Exp. Otorhinolaryngol.* **2020**, *13*, 95–105. [CrossRef] [PubMed]
27. Algren, M.H.; Ekholm, O.; Nielsen, L.; Ersboll, A.K.; Bak, C.K.; Andersen, P.T. Associations between perceived stress, socioeconomic status, and health-risk behaviour in deprived neighbourhoods in Denmark: A cross-sectional study. *BMC Public Health* **2018**, *18*, 250. [CrossRef] [PubMed]
28. Kikuchi, H.; Machida, M.; Nakamura, I.; Saito, R.; Odagiri, Y.; Kojima, T.; Watanabe, H.; Inoue, S. Development of severe psychological distress among low-income individuals during the COVID-19 pandemic: Longitudinal study. *BJPsych Open* **2021**, *7*, e50. [CrossRef]
29. Segerstrom, S.C.; Miller, G.E. Psychological stress and the human immune system: A meta-analytic study of 30 years of inquiry. *Psychol. Bull.* **2004**, *130*, 601–630. [CrossRef]
30. Raifman, M.A.; Raifman, J.R. Disparities in the Population at Risk of Severe Illness From COVID-19 by Race/Ethnicity and Income. *Am. J. Prev. Med.* **2020**, *59*, 137–139. [CrossRef]
31. Sese, L.; Nguyen, Y.; Giroux Leprieur, E.; Annesi-Maesano, I.; Cavalin, C.; Goupil de Bouille, J.; Demestier, L.; Dhote, R.; Tandjaoui-Lambiotte, Y.; Bauvois, A.; et al. Impact of socioeconomic status in patients hospitalised for COVID-19 in the Greater Paris area. *Eur. Respir. J.* **2020**, *56*, 2002364. [CrossRef]
32. Uninsurance noMUCotCo. Care without Coverage: Too Little, too Late. 2002. Available online: https://www.ncbi.nlm.nih.gov/books/NBK220636/ (accessed on 2 June 2021).
33. Mc Williams, J.M.; Zaslavsky, A.M.; Meara, E.; Ayanian, J.Z. Health insurance coverage and mortality among the near-elderly. *Health Aff.* **2004**, *23*, 223–233. [CrossRef]
34. Webb Hooper, M.; Napoles, A.M.; Perez-Stable, E.J. COVID-19 and Racial/Ethnic Disparities. *JAMA* **2020**, *323*, 2466–2467. [CrossRef]

Journal of
Clinical Medicine

Article

Direct and Indirect Effects of SARS-CoV-2 Pandemic in Subjects with Familial Hypercholesterolemia: A Single Lipid-Center Real-World Evaluation

Roberto Scicali [1,2,*], Salvatore Piro [1], Viviana Ferrara [1], Stefania Di Mauro [1], Agnese Filippello [1], Alessandra Scamporrino [1], Marcello Romano [2], Francesco Purrello [1] and Antonino Di Pino [1]

[1] Department of Clinical and Experimental Medicine, University of Catania, 95100 Catania, Italy; spiro@unict.it (S.P.); vivi.fer@hotmail.it (V.F.); 8stefaniadimauro6@gmail.com (S.D.M.); agnese.filippello@gmail.com (A.F.); alessandraska@hotmail.com (A.S.); francesco.purrello@unict.it (F.P.); nino_dipino@hotmail.com (A.D.P.)

[2] Geriatric Unit, Garibaldi Hospital, 95100 Catania, Italy; marcelloromanoct@gmail.com

* Correspondence: robertoscicali@gmail.com; Tel.: +39-0957593945; Fax: +39-0957598123

Abstract: We evaluated the impact of direct and indirect effects of SARS-CoV-2 infection in subjects with familial hypercholesterolemia (FH). In this observational, retrospective study, 260 FH subjects participated in a telephone survey concerning lipid profile values, lipidologist and cardiologist consultations and vascular imaging evaluation during the 12 months before and after the Italian lockdown. The direct effect was defined as SARS-CoV-2 infection; the indirect effect was defined as the difference in one of the parameters evaluated by the telephone survey before and after lockdown. Among FH subjects, the percentage of the lipid profile evaluation was lower after lockdown than before lockdown (56.5% vs. 100.0%, $p < 0.01$), HDL-C was significantly reduced (47.78 ± 10.12 vs. 53.2 ± 10.38 mg/dL, $p < 0.05$) and a significant increase in non-HDL-C was found (117.24 ± 18.83 vs. 133.09 ± 19.01 mg/dL, $p < 0.05$). The proportions of lipidologist and/or cardiologist consultations and/or vascular imaging were lower after lockdown than before lockdown (for lipidologist consultation 33.5% vs. 100.0%, $p < 0.001$; for cardiologist consultation 22.3% vs. 60.8%, $p < 0.01$; for vascular imaging 19.6% vs. 100.0%, $p < 0.001$); the main cause of missed lipid profile analysis and/or healthcare consultation was the fear of SARS-CoV-2 contagion. The percentage of FH subjects affected by SARS-CoV-2 was 7.3%. In conclusion, a lower percentage of FH subjects underwent a lipid profile analysis, lipidologist and cardiologist consultations and vascular imaging evaluation after SARS-CoV-2 Italian lockdown.

Keywords: SARS-CoV-2 pandemic; familial hypercholesterolemia; lipid-lowering therapy; healthcare system; cardiovascular risk

Citation: Scicali, R.; Piro, S.; Ferrara, V.; Di Mauro, S.; Filippello, A.; Scamporrino, A.; Romano, M.; Purrello, F.; Di Pino, A. Direct and Indirect Effects of SARS-CoV-2 Pandemic in Subjects with Familial Hypercholesterolemia: A Single Lipid-Center Real-World Evaluation. *J. Clin. Med.* **2021**, *10*, 4363. https://doi.org/10.3390/jcm10194363

Academic Editors: Giorgio I. Russo and Maciej Banach

Received: 31 August 2021
Accepted: 22 September 2021
Published: 24 September 2021

Publisher's Note: MDPI stays neutral with regard to jurisdictional claims in published maps and institutional affiliations.

1. Introduction

Since December 2019, the severe acute respiratory syndrome coronavirus 2 (SARS-CoV-2) pandemic has affected more than 190,000,000 subjects and caused more than 4,000,000 deaths worldwide [1]. The clinical manifestations of SARS-CoV-2 infection broadly differ among the affected subjects; about half of the infected subjects remain asymptomatic, the majority of the symptomatic subjects experience influenza-like symptoms and 10–15% of these develop a severe disease (COVID-19) characterized by a wide clinical scenario from pneumonia to acute respiratory distress syndrome (ARDS) and disseminated intravascular coagulation [2].

Other than the respiratory tract, COVID-19 can also affect the cardiovascular system. In fact, several mechanisms of SARS-CoV-2 heart injury have been hypothesized: direct myocardial damage by binding the SARS-CoV-2 spike glycoprotein to angiotensin-converting enzyme 2, cardiac inflammation in the context of cytokine release syndrome (cytokine storm) caused by progression of COVID-19, increased myocardial distress in the context of ARDS and coronary plaque rupture due to increased endothelial shear stress [3].

Beyond the reported direct damage of SARS-CoV-2 infection, increasing attention has been focused on the indirect effects of the SARS-CoV-2 pandemic because of the healthcare public system restructuring; in particular, a substantial reduction in hospital admissions for acute coronary syndromes (ACS) during the SARS-CoV-2 pandemic was shown related to the national lockdown in Italy, and this could be explained by increasing fear of in-hospital contagion, an emergency department overload and the healthcare structure remodeling [4]. In this context, the reduced cardiovascular screening may be deleterious in subjects at high cardiovascular risk such as those with familial hypercholesterolemia (FH), which is the most frequent monogenic disorder characterized by a lifelong elevation of low-density lipoprotein cholesterol (LDL-C) and early atherosclerotic cardiovascular disease (ASCVD) [5]. Thus, the delay of clinical and/or genetic diagnosis and the deferred lipid-lowering therapy optimization could promote an increase in LDL-C burden strongly associated with atherosclerotic injury progression [6–8].

In this study, we evaluated the direct and indirect effects of the SARS-CoV-2 pandemic in a cohort of FH subjects.

2. Methods

2.1. Study Design and Population

This was a retrospective, observational study involving patients aged over 18 years with a genetically confirmed FH diagnosis [9] and enrolled from the Lipid Centre of the University Hospital of Catania, Italy, from 4 June 2021 to 9 August 2021. All participants had a telephone survey concerning their lipid profile values (total cholesterol, high-density lipoprotein cholesterol (HDL-C), triglycerides (TG), LDL-C), lipidologist and cardiologist consultations, vascular imaging evaluation and lipid-lowering therapy adherence in the 12 months before and after the Italian lockdown (9 March–3 June 2020); moreover, all participants confirmed or not the SARS-CoV-2 infection from 9 March 2020 to 12 months after the end of the Italian lockdown (3 June 2020). Vascular imaging was defined by carotid and/or femoral ultrasound evaluation. Statin therapy was divided into three categories according to the efficacy of LDL-C reduction: high intensity ($\geq$50% LDL-C reduction, rosuvastatin 20–40 mg/day or atorvastatin 40–80 mg/day), moderate intensity (30–50% LDL-C reduction, rosuvastatin 5–10 mg/day, atorvastatin 10–20 mg/day, simvastatin 20–40-80 mg/day, pravastatin 40 mg/day, fluvastatin 80 mg/day, lovastatin 40 mg/day) or low intensity (<30% LDL-C reduction, simvastatin 10 mg, pravastatin 20 mg/day, fluvastatin 40 mg/day, lovastatin 20 mg/day). Type 2 diabetes and arterial hypertension were defined as the daily intake of glucose-lowering medication and antihypertensive drugs, respectively. ASCVD and LDL-C targets were defined as previously described [10]. Long-wait consultation was defined as >3 months. Hospitalizations for COVID or other comorbidities were also reported. The SARS-CoV-2 pandemic direct effect was defined as the virus-related infection from 9 March 2020 to 12 months after the end of the Italian lockdown. The SARS-CoV-2 pandemic indirect effect was defined as the difference in one of the following evaluated parameters in the 12 months before and after the Italian lockdown: lipid profile analysis, lipidologist and cardiologist consultations and vascular imaging evaluation.

2.2. Statistical Analysis

The distributional characteristics of each variable, including normality, were assessed by the Kolmogorov–Smirnov test. Data are reported as mean $\pm$ standard deviation (SD) for continuous parametric parameters, median (interquartile range (IQR)) for continuous nonparametric variables and frequency (percentage) for categorical variables. When necessary, the continuous nonparametric variable "TG" was logarithmically transformed to reduce skewness. To test differences in clinical and biochemical characteristics of the study population before and after Italian lockdown, we used Student's t-test. The χ^2 test was used for categorical variables. All statistical analyses were performed using IBM SPSS Statistics for Windows version 23. For all tests, $p < 0.05$ was considered significant.

The study was approved by the local ethics committee (prot. number 46/19) in accordance with the ethical standards of the institutional and national research committees and with the 1964 Declaration of Helsinki and its later amendments or comparable ethical standards.

Informed consent was obtained from each subject enrolled in the study.

3. Results

In total, 292 genetically confirmed FH subjects were evaluated; of these, 30 subjects did not satisfy the inclusion criteria and 2 subjects declined. Finally, 260 FH subjects participated in the study (Figure 1).

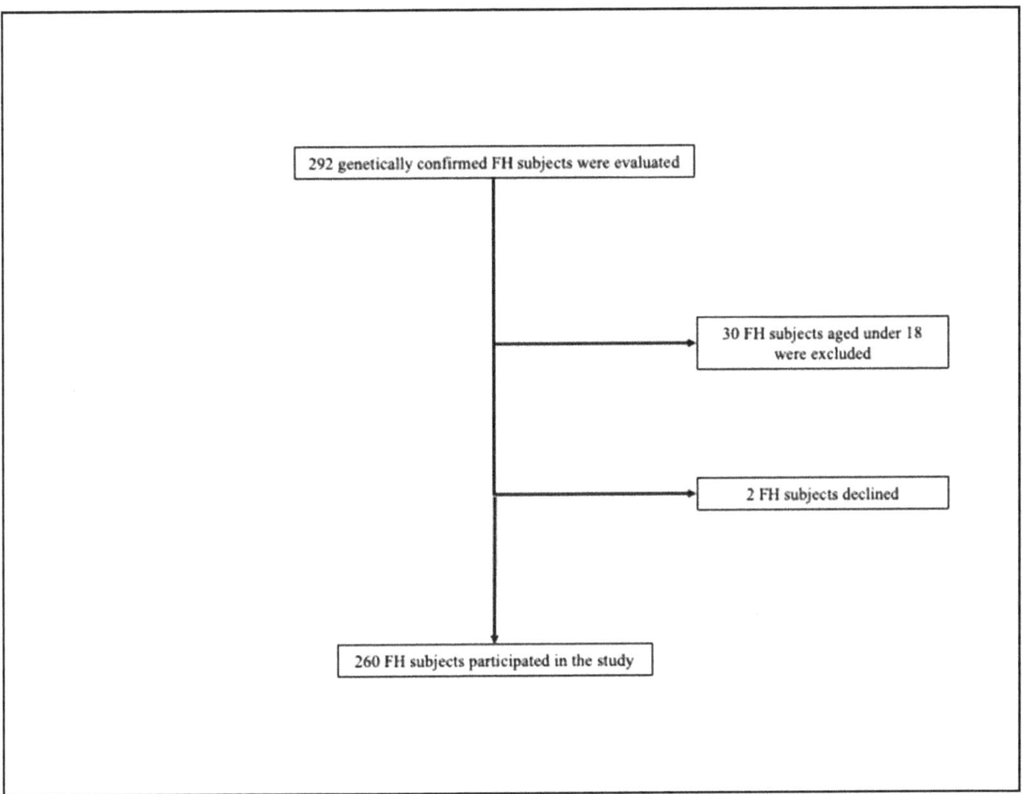

Figure 1. Enrollment flowchart of the study population. FH = familial hypercholesterolemia.

Table 1 shows the baseline characteristics of the study population; 49.6% of FH subjects were males, and the percentage of subjects with a history of ASCVD was 30.8%. The majority of FH subjects exhibited a pathogenic variant in the LDL receptor (LDLR), and 97.7% of subjects were heterozygotes; three subjects were double heterozygotes, two subjects were compound heterozygotes and one subject was homozygote. Concerning the presence of cardiovascular risk factors, the percentage of diabetic FH subjects was 2.3%, 27.7% of subjects were hypertensive and 22.7% of subjects were smokers; the proportion of FH subjects with at least two of the mentioned risk factors was 13.1%. Concerning lipid-lowering treatments, the majority of FH subjects were on statins; in particular, 73.3% of subjects took high-intensity statins, 24.6% of subjects were on moderate-intensity statins and only 1.9% of subjects were statin-intolerant. Furthermore, the percentage of FH subjects on ezetimibe was 86.5%, and 23.8% of subjects were on PCSK9-i therapy; finally, the proportion of subjects on statin and ezetimibe and PCSK9-i was 21.9%.

Table 1. Baseline characteristics of the study population.

	FH (*n* = 260)
Demographic Characteristics Age, years	49.4 ± 6.22
Men, *n* (%)	129 (49.6)
ASCVD, *n* (%)	80 (30.8)
Body mass index, kg/m^2	25.3 ± 2.24
FH Genotype	
Pathogenic variants, *n* (%)	267 (100.0)
LDLR, *n* (%)	261 (97.7)
ApoB, *n* (%)	4 (1.5)
PCSK9, *n* (%)	1 (0.4)
ApoE, *n* (%)	1 (0.4)
FH Phenotype	
Heterozygous, *n* (%)	254 (97.7)
Double heterozygous, *n* (%)	3 (1.1)
Compound heterozygous, *n* (%)	2 (0.8)
Homozygous, *n* (%)	1 (0.4)
Pretreated Lipid Profile TC, mg/dL	362.38 ± 19.48
HDL-C, mg/dL	51.38 ± 10.5
TG, mg/dL	96.5 (71.5–115.5)
LDL-C, mg/dL	257.53 ± 18.15
Non-HDL-C, mg/dL	301.51 ± 19.12
Risk Factors Type 2 diabetes, *n* (%)	6 (2.3)
Hypertension, *n* (%)	72 (27.7)
Smokers, *n* (%)	59 (22.7)
≥2 risk factors, *n* (%)	34 (13.1)
Treatments	
High-intensity statin, *n* (%)	191 (73.5)
Moderate-intensity statin, *n* (%)	64 (24.6)
Low-intensity statin, *n* (%)	-
Statin intolerant, *n* (%)	5 (1.9)
Ezetimibe, *n* (%)	225 (86.5)
PCSK9 inhibitor, *n* (%)	62 (23.8)
Statin plus ezetimibe, *n* (%)	195 (75.0)
Statin plus ezetimibe plus PCSK9 inhibitor, *n* (%)	57 (21.9)
Antiplatelet therapy, *n* (%)	80 (30.8)

Data are presented as mean ± standard deviation, percentages or median (interquartile range). FH = familial hypercholesterolemia, ASCVD = atherosclerotic cardiovascular disease, LDLR = low-density lipoprotein receptor, ApoB = apolipoprotein B, PCSK9 = proprotein convertase subtilisin-kexin type 9, ApoE = apolipoprotein E, TC = total cholesterol, HDL-C = high-density lipoprotein cholesterol, TG = triglycerides, LDL-C = low-density lipoprotein cholesterol.

The direct and indirect effects of SARS-CoV-2 are reported in Table 2. Among FH subjects, the percentage of the lipid profile evaluation was lower after lockdown than before lockdown (56.5% vs. 100.0%, $p < 0.01$); moreover, HDL-C was significantly reduced after lockdown compared to before lockdown (47.78 ± 10.12 vs. 53.2 ± 10.38 mg/dL, $p < 0.05$), and a significant increase in non-HDL-C was found after lockdown compared to before lockdown (117.24 ± 18.83 vs. 133.09 ± 19.01 mg/dL, $p < 0.05$).

Table 2. Direct and indirect effects of SARS-CoV-2 pandemic in the study population.

	FH (*n* = 260) before Lockdown	FH (*n* = 260) after Lockdown	*p* Value
Indirect Effect			
Lipid Profile, *n* (%)	260 (100.0)	147 (56.5)	<0.01
TC, mg/dL *	169.61 ± 18.75	177.83 ± 18.91	0.43
HDL-C, mg/dL *	53.2 ± 10.38	47.78 ± 10.12	<0.05
TG, mg/dL *	90.5 (68.25–114.5)	97.5 (70.5–121.25)	0.11
LDL-C, mg/dL *	103.13 ± 18.02	111.32 ± 18.14	0.25
Non-HDL-C, mg/dL *	117.24 ± 18.83	133.09 ± 19.01	<0.05
LDL-C target, *n* (%) *	105 (40.4)	81 (31.2)	0.09
Lipidologist consultation, *n* (%)	260 (100.0)	87 (33.5)	<0.001
Cardiologist consultation, *n* (%)	158 (60.8)	58 (22.3)	<0.01
Vascular imaging, *n* (%)	260 (100.0)	51 (19.6)	<0.001
Cause of Indirect Effect			
Contagion fear, *n* (%)	-	218 (83.8)	-
Long-wait consultation, *n* (%)	-	42 (16.2)	-
Direct Effect			
SARS-CoV-2 infection, *n* (%)	-	19 (7.3)	-
Hospitalization			
COVID-19, *n* (%)	-	-	-
Other causes, *n* (%)	-	-	-

Data are presented as mean ± standard deviation, percentages, or median (interquartile range). FH = familial hypercholesterolemia, TC = total cholesterol, HDL-C = high-density lipoprotein cholesterol, TG = triglycerides, LDL-C = low-density lipoprotein cholesterol, LLT = lipid-lowering therapy, SARS-CoV-2 = severe acute respiratory syndrome coronavirus 2, COVID = coronavirus disease. * Student's t-test was performed in subjects for whom the lipid profile was evaluated before and after lockdown.

The proportion of FH subjects who had lipidologist and/or cardiologist consultations and/or vascular imaging was lower after lockdown than before lockdown (for lipidologist consultation 33.5% vs. 100.0%, $p < 0.001$; for cardiologist consultation 22.3% vs. 60.8%, $p < 0.01$; for vascular imaging 19.6% vs. 100.0%, $p < 0.001$) (Figure 2); the main cause of missed lipid profile analysis and/or healthcare consultations was the fear of contagion. Finally, the percentage of FH subjects affected by SARS-CoV-2 was 7.3%, and none of them required hospitalization.

As concerns the FH subjects who reported having contracted SARS-CoV-2 infection (Table 3), the mean age was 58.7 ± 5.18, 52.6% of subjects were males and the proportion of subjects with a history of ASCVD was 78.9%. While the majority of SARS-CoV-2-affected FH subjects were on intensive lipid-lowering therapies, only 42.1% of subjects achieved the LDL-C target according to the European Society of Cardiology/European Atherosclerosis Society Guidelines 2019 for the management of dyslipidemias. Concerning the cardiovascular risk factors, 15.8% of FH subjects were diabetics, 84.2% of subjects were hypertensive and 36.8% of subjects were smokers; the percentage of FH subjects with at least two of the mentioned risk factors was 52.6%. Finally, the majority of FH subjects were on high-intensity statins, all subjects took ezetimibe and the proportion of subjects on statins plus ezetimibe plus PCSK9-i was 63.2%.

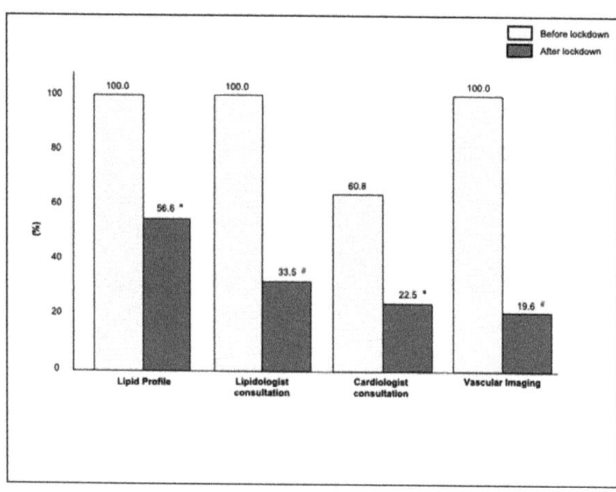

Figure 2. Percentages of lipid profile analysis, lipidologist and cardiologist consultations and vascular imaging evaluation in the study population. * p value < 0.01 vs. before lockdown, # p value < 0.001 vs. before lockdown.

Table 3. Characteristics of FH subjects affected by SARS-CoV-2.

	SARS-CoV-2 FH (n = 19)
Demographic Characteristics	
Age, years	58.7 ± 5.18
Male, n (%)	10 (52.6)
Body mass index, kg/m^2	26.1 ± 1.52
ASCVD, n (%)	15 (78.9)
FH Phenotype	
Heterozygote, n (%)	15 (78.8)
Double heterozygote, n (%)	1 (5.3)
Compound heterozygote, n (%)	2 (10.6)
Homozygote, n (%)	1 (5.3)
Lipid Profile Before Lockdown	
TC, mg/dL	162.45 ± 10.24
HDL-C, mg/dL	49.8 ± 10.13
TG, mg/dL	97.25 (66.0–113.5)
LDL-C, mg/dL	93.34 ± 10.11
Non-HDL-C, mg/dL	113.36 ± 10.43
LDL-C target, n (%)	8 (42.1)
Risk Factors	
Type 2 diabetes, n (%)	3 (15.8)
Hypertension, n (%)	16 (84.2)
Smokers, n (%)	7 (36.8)
≥2 risk factors, n (%)	10 (52.6)
Treatments	
High-intensity statin, n (%)	17 (89.5)
Moderate-intensity statin, n (%)	2 (10.5)
Low-intensity statin, n (%)	-
Statin intolerant, n (%)	-
Ezetimibe, n (%)	19 (100)
PCSK9 inhibitor, n (%)	12 (63.2)
Statin + ezetimibe + PCSK9 inhibitor, n (%)	12 (63.2)
Antiplatelet therapy, n (%)	15 (78.9)

Data are presented as mean ± standard deviation, percentages or median (interquartile range). FH = familial hypercholesterolemia, ASCVD = atherosclerotic cardiovascular disease, TC = total cholesterol, HDL-C = high-density lipoprotein cholesterol, TG = triglycerides, LDL-C = low-density lipoprotein cholesterol, PCSK9 = proprotein convertase subtilisin-kexin type 9.

4. Discussion

Over the last year, increasing attention has been focused on the direct effects of the SARS-CoV-2 pandemic such as the prevalence of infection, COVID-19, hospitalization and death and its indirect effect related to the impact of SARS-CoV-2 pandemic on the healthcare system. In this retrospective observational study, we evaluated the impact of direct and indirect effects of SARS-CoV-2 pandemic in a cohort of subjects at high cardiovascular risk; to the best of our knowledge, this is the first study focusing on SARS-CoV-2 pandemic impact in this population. We found that a lower percentage of FH subjects underwent lipid profile evaluation after the SARS-CoV-2 Italian lockdown; furthermore, a reduction in HDL-C and an increase in non-HDL-C were observed in FH subjects after lockdown. In this context, a hypothetical explanation of these findings could be a dysregulated lifestyle including reduced physical activity and a high-fat diet during the SARS-CoV-2 pandemic; in line with this hypothesis, previous findings showed that reduced physical activity and an increase in BMI were two main effects of the SARS-CoV-2 lockdown [11,12].

In our study, we found that a lower proportion of FH subjects received lipidologist and cardiologist consultations and vascular imaging evaluation after the SARS-CoV-2 lockdown; the main explanation of these findings obtained from FH subjects by telephone survey was the fear of SARS-CoV-2 contagion. This finding was in line with two previous findings that evaluated the impact of SARS-CoV-2 infection on the healthcare system. In fact, Cori et al. showed in the EPICOVID19 web-based Italian survey that 65% of subjects reported fear of SARS-CoV-2 contagion for themselves and family members [13]; moreover, Amorim et al. reported that the admission of patients with ST-elevation myocardial infarction was significantly reduced in the emergency department during SARS-CoV-2 lockdown [14]. Taking into these findings, our study highlighted that the indirect effects of the SARS-CoV-2 pandemic could be deleterious in the cardiovascular risk management of FH subjects; future prospective studies are needed to evaluate the prognostic role of our findings.

In our study, the percentage of FH subjects with SARS-CoV-2 infection was 7.3%, in line with Italian SARS-CoV-2 prevalence [15]; in this context, it could be hypothesized that the risk of SARS-CoV-2 infection was similar between FH subjects and the general population. Moreover, we found that SARS-CoV-2-affected FH subjects had a BMI and age over the mean of the study population, and the majority of them had a prior ASCVD. Furthermore, SARS-CoV-2-affected FH subjects had a before-lockdown HDL-C under the mean of the study population; thus, it could be hypothesized that a low HDL-C level could increase the risk of SARS-CoV-2 infection. In line with this hypothesis, Hilser et al. found that a 10 mg/dL increase in HDL-C or apolipoprotein AI was associated with a 10% reduction in risk of SARS-CoV-2 infection [16]. Although the majority of SARS-CoV-2-affected FH subjects were on intensive lipid-lowering therapy, only 40% of them achieved the LDL-C target according to the European Society of Cardiology/European Atherosclerosis Society Guidelines 2019 for the management of dyslipidemias; moreover, more than 50% of SARS-CoV-2-affected FH subjects had two or more cardiovascular risk factors. Thus, it could be hypothesized that an LDL-C beyond the recommended targets in concomitance with other cardiovascular risk factors could increase the risk of SARS-CoV-2 infection in FH subjects. In line with this hypothesis, Lusignan et al. showed in the Oxford Royal College of General Practitioners (RCGP) Research and Surveillance Centre primary care network that subjects with SARS-CoV-2 infection had several cardiovascular risk factors and the presence of diabetes and/or smoking and/or arterial hypertension increased the risk of SARS-CoV-2 infection [17]. Finally, in our study, none of the affected FH subjects required hospitalization; future studies in larger cohorts of FH subjects are needed to confirm and explain this preliminary finding. However, previous findings showed that among subjects with SARS-CoV-2 infection requiring hospitalization, statin users were associated with lower mortality than non-statin users [18–20]. Taking these findings into consideration, a possible hypothesis could be that subjects with a long duration of statin therapy, such as FH subjects, could be characterized by a reduced need for hospitalization [21].

There are several limitations to our study. First, this was a retrospective, observational study, and lifestyle evaluation and lipid-lowering therapy adherence were not reported; future prospective studies are needed to correctly evaluate these parameters. Moreover, the study population size was relatively small; for this reason, our preliminary findings should be confirmed in a larger cohort of FH subjects. Finally, a possible pathophysiological link of the atherosclerotic injury in SARS-CoV-2 subjects and FH subjects has not been evaluated; future prospective studies are needed to evaluate this feature.

In conclusion, a lower percentage of FH subjects underwent a lipid profile analysis, lipidologist and cardiologist consultations and vascular imaging evaluation after SARS-CoV-2 lockdown; moreover, reduced HDL-C and increased non-HDL-C were observed in FH subjects after SARS-CoV-2 lockdown, Finally, SARS-CoV-2-affected FH subjects exhibited an LDL-C beyond the recommended targets in concomitance with other cardiovascular risk factors; future prospective studies in a larger cohort of FH subjects are needed to confirm these preliminary findings.

Author Contributions: Conceptualization, R.S., S.P., M.R., F.P. and A.D.P.; Data curation, R.S., V.F., S.D.M., A.F., A.S. and A.D.P.; Formal analysis, R.S.; Investigation, V.F., S.D.M., A.F. and A.S.; Methodology, R.S., V.F. and A.D.P.; Supervision, S.P., F.P. and A.D.P.; Validation, S.P., M.R., F.P. and A.D.P.; Writing—original draft, R.S.; Writing—review & editing, A.D.P. All authors have read and agreed to the published version of the manuscript.

Funding: This research received no external funding.

Institutional Review Board Statement: The study was conducted according to the guidelines of the Declaration of Helsinki, and approved by the Ethics Committee CATANIA 2 of Garibaldi Hospital (protocol code 46/19 and date of approval 07-10-2015).

Informed Consent Statement: Informed consent was obtained from all subjects involved in the study.

Acknowledgments: Genetic analysis was carried out within the Lipigen study, an initiative of the SISA Foundation. The authors wish to thank the Department of Clinical and Experimental Medicine for financial support in the context of the 2016/2018 Department Research Plan of the University of Catania (project #A). The authors wish to thank the Scientific Bureau of the University of Catania for language support.

Conflicts of Interest: The authors declare no conflict of interest.

References

1. World Health Organization Coronavirus Disease (COVID-19). Available online: https://www.who.int/emergencies/diseases/novel-coronavirus-2019 (accessed on 20 July 2021).
2. Guan, W.; Ni, Z.; Hu, Y.; Liang, W.; Ou, C.; He, J.; Liu, L.; Shan, H.; Lei, C.; Hui, D.S.C.; et al. Clinical Characteristics of Coronavirus Disease 2019 in China. *N. Engl. J. Med.* **2020**, *382*, 1708–1720. [CrossRef] [PubMed]
3. Madjid, M.; Safavi-Naeini, P.; Solomon, S.D.; Vardeny, O. Potential Effects of Coronaviruses on the Cardiovascular System: A Review. *JAMA Cardiol.* **2020**, *5*, 831–840. [CrossRef] [PubMed]
4. Stefanini, G.G.; Azzolini, E.; Condorelli, G. Critical Organizational Issues for Cardiologists in the COVID-19 Outbreak. *Circulation* **2020**, *141*, 1597–1599. [CrossRef] [PubMed]
5. Scicali, R.; Di Pino, A.; Platania, R.; Purrazzo, G.; Ferrara, V.; Giannone, A.; Urbano, F.; Filippello, A.; Rapisarda, V.; Farruggia, E.; et al. Detecting familial hypercholesterolemia by serum lipid profile screening in a hospital setting: Clinical, genetic and atherosclerotic burden profile. *Nutr. Metab. Cardiovasc. Dis.* **2018**, *28*, 35–43. [CrossRef] [PubMed]
6. Mandraffino, G.; Scicali, R.; Rodríguez-Carrio, J.; Savarino, F.; Mamone, F.; Scuruchi, M.; Cinquegrani, M.; Imbalzano, E.; Di Pino, A.; Piro, S.; et al. Arterial stiffness improvement after adding on PCSK9 inhibitors or ezetimibe to high-intensity statins in patients with familial hypercholesterolemia: A Two–Lipid Center Real-World Experience. *J. Clin. Lipidol.* **2020**, *14*, 231–240. [CrossRef] [PubMed]
7. Basili, S.; Loffredo, L.; Pastori, D.; Proieti, M.; Farcomeni, A.; Vesti, A.R.; Pignatelli, P.; Davì, G.; Hiatt, W.R.; Lip, G.Y.H.; et al. Carotid plaque detection improves the predictve value of CHA2DS2-VASc score in patients with non-valvular atrial fibrillation: The ARAPACIS Study. *Int. J. Cardiol.* **2017**, *231*, 143–149. [CrossRef] [PubMed]
8. Banach, M.; Penson, P.E.; Fras, Z.; Vrablik, M.; Pella, D.; Reiner, Ž.; Nabavi, S.M.; Sahebkar, A.; Kayikcioglu, M.; Daccord, M.; et al. Brief recommendations on the management of adult patients with familial hypercholesterolemia during the COVID-19 pandemic. *Pharmacol. Res.* **2020**, *158*, 104891. [CrossRef] [PubMed]

9. Pirillo, A.; Garlaschelli, K.; Arca, M.; Averna, M.; Bertolini, S.; Calandra, S.; Tarugi, P.; Catapano, A.L.; Arca, M.; Averna, M.; et al. Spectrum of mutations in Italian patients with familial hypercholesterolemia: New results from the LIPIGEN study. *Atheroscler. Suppl.* **2017**, *29*, 17–24. [CrossRef] [PubMed]

10. Scicali, R.; Russo, G.I.; Di Mauro, M.; Manuele, F.; Di Marco, G.; Di Pino, A.; Ferrara, V.; Rabuazzo, A.M.; Piro, S.; Morgia, G.; et al. Analysis of Arterial Stiffness and Sexual Function after Adding on PCSK9 Inhibitor Treatment in Male Patients with Familial Hypercholesterolemia: A Single Lipid Center Real-World Experience. *J. Clin. Med.* **2020**, *9*, 3597. [CrossRef] [PubMed]

11. Puccinelli, P.J.; da Costa, T.S.; Seffrin, A.; de Lira, C.A.B.; Vancini, R.L.; Nikolaidis, P.T.; Knechtle, B.; Rosemann, T.; Hill, L.; Andrade, M.S. Reduced level of physical activity during COVID-19 pandemic is associated with depression and anxiety levels: An internet-based survey. *BMC Public Healhth* **2021**, *21*, 1–11. [CrossRef]

12. Bakaloudi, D.R.; Barazzoni, R.; Bischoff, S.C.; Breda, J.; Wickramasinghe, K.; Chourdakis, M. Impact of the first COVID-19 lockdown on body weight: A combined systematic review and a meta-analysis. *Clin. Nutr.* **2021**, in press. [CrossRef]

13. Cori, L.; Curzio, O.; Adorni, F.; Prinelli, F.; Noale, M.; Trevisan, C.; Fortunato, L.; Giacomelli, A.; Bianchi, F. Fear of COVID-19 for Individuals and Family Members: Indications from the National Cross-Sectional Study of the EPICOVID19 Web-Based Survey. *Int. J. Environ. Res. Public Health* **2021**, *18*, 3248. [CrossRef] [PubMed]

14. Pessoa-Amorim, G.; Camm, C.F.; Gajendragadkar, P.; De Maria, G.L.; Arsac, C.; Laroche, C.; Zamorano, J.L.; Weidinger, F.; Achenbach, S.; Maggioni, A.P.; et al. Admission of patients with STEMI since the outbreak of the COVID-19 pandemic: A survey by the European Society of Cardiology. *Eur. Heart J. Qual. Care Clin. Outcomes* **2020**, *6*, 210–216. [CrossRef] [PubMed]

15. WHO Coronavirus (COVID-19) Dashboard | WHO Coronavirus (COVID-19) Dashboard with Vaccination Data. Available online: https://covid19.who.int/ (accessed on 13 July 2021).

16. Hilser, J.R.; Han, Y.; Biswas, S.; Gukasyan, J.; Cai, Z.; Zhu, R.; Tang, W.H.W.; Deb, A.; Lusis, A.J.; Hartiala, J.A.; et al. Association of serum HDL-cholesterol and apolipoprotein A1 levels with risk of severe SARS-CoV-2 infection. *J. Lipid Res.* **2021**, *62*, 100061. [CrossRef] [PubMed]

17. de Lusignan, S.; Dorward, J.; Correa, A.; Jones, N.; Akinyemi, O.; Amirthalingam, G.; Andrews, N.; Byford, R.; Dabrera, G.; Elliot, A.; et al. Risk factors for SARS-CoV-2 among patients in the Oxford Royal College of General Practitioners Research and Surveillance Centre primary care network: A cross-sectional study. *Lancet. Infect. Dis.* **2020**, *20*, 1034. [CrossRef]

18. Chacko, S.R.; DeJoy, R.; Lo, K.B.; Albano, J.; Peterson, E.; Bhargav, R.; Gu, F.; Salacup, G.; Pelayo, J.; Azmaiparashvili, Z.; et al. Association of Pre-Admission Statin Use With Reduced In-Hospital Mortality in COVID-19. *Am. J. Med. Sci.* **2021**, *361*, 725–730. [CrossRef] [PubMed]

19. Masana, L.; Correig, E.; Rodríguez-Borjabad, C.; Anoro, E.; Arroyo, J.A.; Jericó, C.; Pedragosa, A.; Miret, M.; Näf, S.; Pardo, A.; et al. Effect of statin therapy on SARS-CoV-2 infection-related mortality in hospitalized patients. *Eur. Heart J. Cardiovasc. Pharmacother.* **2020**, pvaa128. [CrossRef] [PubMed]

20. Vahedian-Azimi, A.; Mohammadi, S.M.; Beni, F.H.; Banach, M.; Guest, P.C.; Jamialahmadi, T.; Sahebkar, A. Improved COVID-19 ICU admission and mortality outcomes following treatment with statins: A systematic review and meta-analysis. *Arch. Med. Sci.* **2021**, *17*, 579–595. [CrossRef] [PubMed]

21. Radenkovic, D.; Chawla, S.; Pirro, M.; Sahebkar, A.; Banach, M. Cholesterol in Relation to COVID-19: Should We Care about It? *J. Clin. Med.* **2020**, *9*, 1909. [CrossRef] [PubMed]

Journal of
Clinical Medicine

Article

Sexual Behaviour and Fantasies in a Group of Young Italian Cohort

Marina Di Mauro [1], Giorgio Ivan Russo [1,*], Gaia Polloni [2], Camilla Tonioni [3], Daniel Giunti [3], Gianmartin Cito [4], Bruno Giammusso [5], Girolamo Morelli [6], Lorenzo Masieri [4] and Andrea Cocci [4]

1 Urology Section, Department of Surgery, University of Catania, 95100 Catania, Italy; marinadimauro@live.it
2 Centre of Pshycology, Via Cadorna, 22100 Como, Italy; dott.gaiapolloni@gmail.com
3 Centro Integrato di Sessuologia Il Ponte, 50100 Florence, Italy; camilla.tonioni@stud.unifi.it (C.T.); danielgiunti@gmail.com (D.G.)
4 Department of Urology, University of Florence, 50100 Florence, Italy; gianmartin.cito@gmail.com (G.C.); lorenzo.masieri@meyer.it (L.M.); cocci.andrea@gmail.com (A.C.)
5 Morgagni Hospital, 95100 Catania, Italy; bgiammusso@hotmail.it
6 Urology Section, University of Pisa, 56121 Pisa, Italy; girolamomorelli@gmail.com
* Correspondence: giorgioivan1987@gmail.com

Abstract: Over the years, sexual behaviour has changed due to the growing interest in everything related to the sexual sphere. The purpose of the study was to collect information on the sexual habits and behaviours of Italian people of all ages, sexes and sexual orientations and to describe the patterns of sexual behaviour, with the aim of gaining a representative picture of sexuality in Italy, before the COVID-19 pandemic. Participants completed a survey with 99 questions about their sexual habits. In our group first sexual experiences occurred on average around the age of 15, whilst the median age of the first sexual intercourse was 17. The fantasies that most stimulated and excited our group (Likert scale $\geq$ 3) was having sex in public (63.9%), having sex with more than one person at the same time (59.4%), blindfolded sex (64.9%), being tied up (56.3%) and observing a naked person (48.6%). As for pornography, we have shown that 80% of our group watched porn at home, alone or from their smartphones. Our results have several practical implications for the areas of sex education and sexual health. It is necessary to safeguard the health of young people and support them increasing their sexual well-being.

Keywords: sex aid; sexual behavior; pornography; alcohol; erectile dysfunction

Citation: Di Mauro, M.; Russo, G.I.; Polloni, G.; Tonioni, C.; Giunti, D.; Cito, G.; Giammusso, B.; Morelli, G.; Masieri, L.; Cocci, A. Sexual Behaviour and Fantasies in a Group of Young Italian Cohort. *J. Clin. Med.* **2021**, *10*, 4327. https://doi.org/10.3390/jcm10194327

Academic Editor: Du Geon Moon

Received: 9 September 2021
Accepted: 22 September 2021
Published: 23 September 2021

Publisher's Note: MDPI stays neutral with regard to jurisdictional claims in published maps and institutional affiliations.

1. Introduction

Sexual behavior has consistently changed over the years. In fact, there is a growing interest in specific topics that have been considered a stimga in the past, and also physicians should constantly pay attention to patients' preferences [1,2].

Scientific evidences have underlined that the majority of information on sexuality collected through the internet by young people regard explicit messages including and facilitating sexual practices like autoerotism and masturbation [3].

Exploring patterns of current sexual behaviours is important for several reasons, but mainly because the description of behavioural trends can provide an important empirical context for examining the associations between patterns of emerging sexual behaviour and aspects of sexual health and well-being among young people [4]. A recent survey by Herbenick et al. (2020) observed that more frequent past-year pornography use and a greater lifetime range of pornography accesses were significantly associated with engaging in both dominant and target sexual behaviors among all participants [5]. Furthermore, sex aids are also considered tools to help individuals achieving sexual pleasure and can also be particularly helpful for sexual dysfunction [6]. Indeed, sexual fantasies play a major role in influencing later sexual behavior, in reflecting past experiences and these are a core variable in the systematic study of sexual identity and sexuality [7].

Italian society has generally less favourable attitudes towards unions that differ from the traditional wedding [8]. This could be due to the presence of the Catholic Church [9].

Furthermore, although teenagers have the tendency to have their first sexual relationship earlier than had been reported in the past, in Italy it has been observed a decrease of marriage and birthrate [10].

All these considerations may arise some questions about the social background of Italy and the influence on many aspects of sexuality, including internet pornography, sex toys and sexual orientation.

Interestingly, Ross et al. [11] showed that between participants who reported using the internet to retrieve information on sexuality, younger participants displayed higher use of the medium compared to older participants, as well as bisexual men compared to heterosexual men, and males compared to females, respectively, suggesting as internet may facilitate sexual fantasies. Moreover, Daneback and Löfberg [12] suggested that using internet facilitates the expression as well as the engagement of individuals in new experiences, to a degree that would normally be not tolerated.

Finally, The COVID-19-related lockdown has profoundly changed human behaviors and habits, impairing general and psychological well-being with psychosocial consequences on sexual behavior. Jannini et al. demonstrated that anxiety and depression scores were significantly lower in subjects sexually active during lockdown [13]. In particular, sexual activity, and living without partner during lockdown as significantly affecting anxiety and depression scores [13].

Based on all these premises, the scope of web survey was to collect information about sexual habits and behaviours of Italian people of all ages, genders and sexual orientations and to describe patterns of sexual behaviour.

2. Materials and Methods

A quantitative correlational research design was implemented for this study to evaluate the sexual habits in Italian participants in all gender and sexual interest. The study was conducted from 1 June 2019 to 31 December 2019.

Participants were selected through posts on social networks (Instagram and Facebook) and the survey was developed and administered online through Google Forms. Each participant gave the consent to complete the study.

Basic demographic information was collected: gender, age, height, weight, smoking habit, place of residence, sexual orientation, education level, religion and relationship status and duration.

After that, participants completed a survey with questions about their sexual habits. The questions evaluated a variety of aspects: frequency and pleasantness experienced when being involved in various sexual activities (self-stimulation, being masturbated by the partner, masturbate the partner, receiving and giving oral sex, vaginal penetration, receiving and giving anal penetration), sexual satisfaction, frequency of orgasm, stimuli used to get aroused during auto-eroticism, the use of sex toys, pleasantness of various sexual fantasies, pornography use, betrayal, traumatic sexual experiences, stress, contraception, protection against sexually transmitted infections, use of medications or drugs, use of dating apps or sites and sexting.

The survey was conducted in Italian according to the Checklist for Reporting Results of Internet E-Surveys [14].

All the study procedures were carried out in accordance with the Declaration of Helsinki (2013) of the World Medical Association. The survey was anonymous and participants provided their consent to participate.

Statistical Analysis

The qualitative data was tested using the chi-square test or Fisher's exact test, where appropriate, while the continuous variables, presented as median (interquartile range [IQR]), were tested using Mann-Whitney U-Test or Student t test according to their distribution (according to the Kolmogorov-Smirnov test). For all statistical comparisons, significance was considered as $p < 0.05$.

3. Results

3.1. Characteristics of Participants

The median (IQR) age was 20 (18–23) years. Most of the participants enrolled were females, with 7719 (61.3%) individuals, men were 4805 (38.2%), whereas Trans were 20 (0.2%). Participants were stratified by Area of Origin, with 6036 (47.9%) coming from Northern Italy, 2646 (21.0%) from the Center and 3908 (31.0%) from the South and Islands of Italy. The education level was Higher in 7481 (59.4%) of people, with university degrees in 4211 (33.4%). Heterosexual were the most represented participants, with 10,153 (80.6%) people, Homosexual were 234 (1.9%), Bisexual 2087 (16.6%) and Pansexual 83 (0.7%). 2512 (20.0%) participants reported not having a partner, 1325 (10.5%) having occasional partners, 8598 (68.3%) having a stable relationship, 155 (1.2%) having Polyamorous relationships. The median (IQR) duration of relationships was 15 (6–36) months. 12,152 (96.5%) of our participants has no children. The median (IQR) age of the first sexual experiences was 15 (14–17) years whilst the age of the first sexual intercourse was 17 (15–18). Table 1 lists the baseline characteristics of the patients.

Table 1. Basic characteristics of the participants of our study.

Participants, $n = 12{,}590$	
Age, years median (IQR)	20 (18–23)
Height, cm median (IQR)	170 (163–177)
Weight, kg median (IQR)	64 (55–74)
BMI, kg/m^2 median (IQR)	22.1 (20.2–24.7)
Gender, n (%)	
Male	4805 (38.2)
Female	7719 (61.3)
Trans	20 (0.2)
Other	42 (0.3)
Area of Origin, n (%)	
Northern	6036 (47.9)
Center	2646 (21.0)
South and Islands	3908 (31.0)
Education level, n (%)	
Primary education	4 (0.1)
Secondary education	894 (7.1)
Higher education	7481 (59.4)
Universities	4211 (33.4)
Religion, n (%)	
Atheist	4908 (40.9)
Agnostic	1142 (9.5)
Believer	5931 (49.5)
Smoking, n (%)	
Yes	4553 (36.2)
No	8037 (63.8)

Table 1. *Cont.*

Sexual Orientation, n (%)	
Heterosexual	10,153 (80.6)
Homosexual	234 (1.9)
Bisexual	2087 (16.6)
Demi	25 (0.2)
Queer	7 (0.1)
Pansexual	83 (0.7)
Type of relationship, n (%)	
No partner	2512 (20.0)
Occasional partners	1325 (10.5)
Stable relationship	8598 (68.3)
Polyamorous relationship	155 (1.2)
Time of the relationship, months median (IQR)	15 (6–36)
Children, n (%)	
Yes	438 (3.5)
No	12,152 (96.5)
First sexual experiences, age median (IQR)	15 (14–17)
First sexual intercourse, age median (IQR)	17 (15–18)

3.2. Sexual Experience

We questioned responders about their sexual behaviors and we investigated the frequency of each experience by dividing them into "Never", "Few times a year", "About once a month", "About once a week", "Several times a month "," Several times a week "," Several times a day "and" Every day ". These results are shown in Table 2.

Table 2. Sexual experience patterns in the total cohort.

Participants, n = 12,590	
Partner Masturbates You	
Responders, n (%)	
Never	774 (6.1)
Few times a year	571 (4.5)
About once a month	704 (5.6)
About once a week	2701 (21.5)
Several times a month	1745 (13.9)
Several times a week	5258 (41.8)
Several times a day	285 (2.3)
Every day	552 (4.4)

Table 2. *Cont.*

You masturbate your partner	
Responders, *n* (%)	
Never	613 (4.9)
Few times a year	429 (3.4)
About once a month	589 (4.7)
About once a week	2641 (21.0)
Several times a month	1749 (13.9)
Several times a week	5638 (44.8)
Several times a day	336 (2.7)
Every day	595 (4.7)
Partner practices oral sex on you	
Responders, *n* (%)	
Never	1298 (10.3)
Few times a year	918 (7.3)
About once a month	1056 (8.4)
About once a week	2550 (20.3)
Several times a month	2053 (16.3)
Several times a week	4182 (33.2)
Several times a day	190 (1.5)
Every day	343 (2.7)
You practice oral sex on the partner	
Responders, *n* (%)	
Never	967 (7.7)
Few times a year	640 (5.1)
About once a month	874 (6.9)
About once a week	2526 (20.1)
Several times a month	2158 (17.1)
Several times a week	4752 (37.7)
Several times a day	235 (1.9)
Every day	438 (3.5)
Vaginal penetrative intercourse	
Responders, *n* (%)	
Never	1117 (8.9)
Few times a year	446 (3.5)
About once a month	656 (5.2)
About once a week	2450 (19.5)
Several times a month	1620 (12.9)
Several times a week	5366 (42.6)
Several times a day	401 (3.2)
Every day	534 (4.2)

Table 2. *Cont.*

Anal penetrative intercourse (inseritive)	
Responders, *n* (%)	
Never	10,310 (81.9)
Few times a year	1009 (8.0)
About once a month	377 (3.0)
About once a week	230 (1.8)
Several times a month	375 (3.0)
Several times a week	243 (1.9)
Several times a day	22 (0.2)
Every day	24 (0.2)
Anal penetrative intercourse (receptive)	
Responders, *n* (%)	
Never	9896 (78.6)
Few times a year	1356 (10.8)
About once a month	471 (3.7)
About once a week	213 (1.7)
Several times a month	423 (3.4)
Several times a week	193 (1.5)
Several times a day	23 (0.2)
Every day	15 (0.1)
Autoeroticism	
Responders, *n* (%)	
Never	1344 (10.7)
Few times a year	834 (6.6)
About once a month	799 (6.3)
About once a week	1340 (10.6)
Several times a month	1361 (10.8)
Several times a week	4272 (33.9)
Several times a day	789 (6.3)
Every day	1851 (14.7)

3.3. Sex Toys, Sexual Pleasure and Pornography

We asked which sex toys were used during autoeroticism. We investigated what brought pleasure and arousal. The answers were expressed according to the Likert scale, where 1 indicates "not pleasure", 2 indicates "a little pleasure", 3 indicates "enough pleasure", 4 indicates "very pleased" and 5 indicates "maximum pleasure". These results are shown in Supplementary Table S1. Table 3 shows results of sex toys usage, types and use frequency.

Table 3. Sex toys use in the total cohort.

Participants, n = 12,590	
What do you use to get excited during autoeroticism	
Responders, n (%)	
Videos	2802 (22.2)
Sextoys	17 (0.1)
Erotic fantasies	5820 (46.2)
Erotic narrative	980 (7.8)
Erotic images	1573 (12.5)
Nothing	69 (0.5)
I don't practice it	1277 (10.1)
How often do you use sex objects/toys during sexual intercourse	
Responders, n (%)	
I don't have sex	683 (5.4)
Ever	8511 (67.6)
Few times	2502 (19.9)
About half the time	462 (3.7)
Many times	352 (2.8)
Always	80 (0.6)
How often do you use sex objects/Toys during masturbation?	
Responders, n (%)	
I don't have sex	372 (3.0)
Ever	9266 (73.6)
Few times	1731 (13.7)
About half the time	379 (3.0)
Many times	441 (3.5)
Always	401 (3.2)
What kind of sex toys do you use most frequently?	
Responders, n (%)	
Vibrating rings	8 (0.1)
Fruit/vegetables	1048 (8.3)
Cock-rings	114 (0.9)
Sexy underwear	853 (6.8)
Disguise	1013 (8.0)
Fetish objects	1613 (12.8)
Lubricant	95 (0.8)
Dildos	113 (0.9)
Butt plung/anal dilators	7648 (60.8)
Objects for daily use	21 (0.2)
Strap-ons	11 (0.1)
Vibrator	10 (0.1)
Balls	1 (0.0)
Fleshlight (artificial vaginas)	15 (0.1)
I don't use it	26 (0.1)

With our survey, we investigated the use of pornography, by asking in which context, how often and which type of pornographic material they prefer to use (Table 4).

Table 4. Pornography Patterns.

Participants, *n* = 12,590	
In general, in which context do you see pornography most frequently?	
Responders, *n* (%)	
In pairs	503 (4.0)
Alone	10,128 (80.5)
In a group	11 (0.1)
Never	1942 (15.4)
How often do you view online pornography?	
Responders, *n* (%)	
Never	1816 (14.4)
Few times a year	1512 (12.0)
About once a month	1015 (8.1)
About once a week	1468 (11.7)
Several times a month	1457 (11.6)
Several times a week	3501 (27.8)
Many times a day	527 (4.2)
Everyday	1294 (10.3)
What is the most frequent topic of the pornographic material you use?	
Amateur	342 (5.5)
Anal	242 (3.9)
Asian	22 (0.4)
Masturbation	143 (2.3)
Bbw	14 (0.2)
Bdsm	215 (3.5)
Big Ass	26 (0.4)
Big Boobs	85 (1.4)
Big Cock	5 (0.1)
Blonde	19 (0.3)
Bisexual	15 (0.2)
Black	18 (0.3)
Oral Sex	167 (2.7)
Bondage	111 (1.8)
Brazzers	9 (0.1)
Casting	28 (0.5)
Lesbian	734 (11.9)
Classic	106 (1.7)
Compilation	17 (0.3)
Cunnilingus	39 (0.6)
Couple	67 (1.1)
Cowgirl	7 (0.1)
Creampie	65 (1.0)
Cuckold	16 (0.3)

Table 4. *Cont.*

Cumshot	25 (0.4)
Curvy	6 (0.1)
Deepthroat	20 (0.3)
18 years old/Teenagers/Young	572 (9.2)
It Depends	58 (0.9)
Domination	13 (0.2)
Style Stays	12 (0.2)
Double Penetration	45 (0.7)
Doctor	6 (0.1)
Threesome	423 (6.8)
Ebony	12 (0.2)
Erotic	27 (0.4)
Straight	328 (5.3)
Facesitting	14 (0.2)
Fak Taxi Familia/Incesti/Daddy/Stepsister	16 (0.3)
Fantasy	250 (4.0)
Fendom	8 (0.1)
Fetishism	8 (0.1)
Fingering	34 (0.5)
Footjob	13 (0.2)
Gang Bang	34 (0.5)
Fisting	103 (1.7)
Cartons	12 (0.2)
Gay	13 (0.2)
Role Play Games	184 (3.0)
Hardcore	6 (0.1)
Hentai	165 (2.7)
Handjob	95 (1.5)
Several Racies	5 (0.1)
Italian	19 (0.3)
Massage	21 (0.3)
Mature/Milf	66 (1.1)
None In Particular	440 (7.1)
Orgasm	102 (1.6)
Squirting	55 (0.9)
Group Sex	62 (1.0)
Passionate/Romantic	135 (2.2)
Betrayal	14 (0.2)
Transsexual	9 (0.1)
Spanking	17 (0.3)
Public Sex	50 (0.8)
Red Head	16 (0.3)

Table 4. *Cont.*

Pissing	10 (0.2)
Pov	47 (0.8)
Other	110 (1.8)
Where do you view online pornography?	
Responders, *n* (%)	
I don't watch porn	1845 (14.7)
At home	10,721 (85.2)
At work	17 (0.1)
In public	7 (0.1)
Which device do you most frequently view pornography with?	
Responders, *n* (%)	
I don't watch porn	1851 (14.7)
Computers	1031 (8.2)
Video game consoles	18 (0.1)
Smartphones	9294 (73.8)
Tablets	396 (3.1)

At the chi-square test we demonstrated that heterosexuals, homosexuals and bisexuals and were more likely to watch porn more than several times a week (40.9%, 56.4% and 47.4%) respect to Demi (24%), Queer (40.9%) or Pansexual (39.7%) ($p < 0.01$).

The rate of watching lesbian among male, female and other were 22.5%, 77.1% and 0.4% respectively, while for 18 years old/Teenagers/Young category they were 80.9%, 18.9% and 0.2% respectively.

3.4. Contraception and Frequency of Sexual Intercourse

We also investigated the use of contraception, the frequency of sexual intercourse and masturbation and the frequency of reaching orgasm (Supplementary Tables S2 and S3).

3.5. Use of Substances and Dating App

We investigated the possible use of exciting substances, drugs and substances that increase sexual potency (Table 5).

Table 5. Use of Substances among the general cohort.

Participants, *n* = 12,590	
How frequently do you use the following drugs in sexuality?	
Viagra, Cialis, Levitra, Spedra	
Responders, *n* (%)	
Never	12,412 (98.8)
Hardly ever	94 (0.7)
Sometimes	32 (0.3)
Often	8 (0.1)
Always	11 (0.1)

Table 5. *Cont.*

Paroxetine (Daparox/Eutimil), Priligy, Drops/Creams for premature ejaculation	
Responders, *n* (%)	
Never	12,433 (99.0)
Hardly ever	78 (0.6)
Sometimes	25 (0.2)
Often	9 (0.1)
Always	16 (0.1)
Alcohol	
Responders, *n* (%)	
Never	6905 (54.9)
Hardly ever	3207 (25.5)
Sometimes	1851 (14.7)
Often	527 (4.2)
Always	86 (0.7)
Stimulants (Cocaine, Amphetamines etc.)	
Responders, *n* (%)	
Never	12,371 (98.4)
Hardly ever	131 (1.0)
Sometimes	42 (0.3)
Often	15 (0.1)
Always	10 (0.1)
Relaxing (Cannabis etc.)	
Responders, *n* (%)	
Never	9910 (78.8)
Hardly ever	1273 (10.1)
Sometimes	764 (6.1)
Often	431 (3.4)
Always	195 (1.6)
Hallucinogens	
Responders, *n* (%)	
Never	12,464 (99.2)
Hardly ever	68 (0.5)
Sometimes	19 (0.2)
Often	8 (0.1)
Always	9 (0.1)
What reasons push you to use these substances?	
Responders, *n* (%)	
Habitual Use	38 (0.3)
Funny	52 (0.4)
Occasional Use	201 (1.6)
It Like Me	81 (0.6)
Improves Performance	449 (3.6)

Table 5. *Cont.*

Improves Sensations	1400 (11.2)
To Decrease Performance Anxiety	400 (3.2)
To Eliminate The Inhibitor Brakes	1067 (8.6)
For Transgression	537 (4.3)
Relaxation	49 (0.4)
Increase Excitement	16 (0.1)
I Don't Use It	8098 (64.9)
Other Reasons	82 (0.7)
How do you rate sexuality using these substances?	
Responders, *n* (%)	
I Don't Use Any Substance	7976 (63.8)
Not At All Satisfactory	127 (1.0)
Unsatisfactory	418 (3.3)
Quite Satisfactory	2420 (19.3)
Very Satisfying	1570 (12.5)

Male were more likely to use drugs always than other categories (90.9% vs. 1.1%; $p < 0.01$) and similar heterosexuals (81.8% vs. 9.2%; $p < 0.01$).

Finally, we investigated the use of social networks or dating sites to find partners with whom to have sex, and we asked what type of material was exchanged on these platforms. The results are indicated in Supplementary Table S4.

4. Discussion

Our survey has investigated contemporary sexual behaviour in Italy before COVID-19 pandemic in most of its forms, taking into consideration any gender and sexual orientations. When it comes to sex, there is always plenty of curiosity, but, at the same time, often reticence and embarrassment.

Very often, the interest in the sexual habits of the healthy general population is not properly studied. Most of the studies in the literature, in fact, focus on various pathologies of interest and on specific clinical outcomes. However, it is also important to have information on the sexual habits of the healthy general population in order to be able to establish future educational measures but also to provide data on the potential economic impact of the world regarding "sex". Furthermore, considerable interest in the psychological aspects of sexual dysfunctions is also emerging from some international guidelines [15].

Given that the pleasure that sexual pleasure is a fundamental component of sexual health, devices designed to enhance and diversify sexual pleasure could be particularly useful in clinical practice. Despite their growing popularity and widespread use in various biopsychosocial circumstances, many taboos still seem to exist, as indicated by the paucity of scientific literature on the prevalence, application and effectiveness of sexual devices for therapeutic use [16].

Interestingly, compared to our European fellows, the use of sex toys in Italy is not widespread. In fact, a study of Döring et al. (2019) showed that in Germany about 50% of the respondents reported using sex toys both when masturbating and in presence of a partner [1]. Instead, our survey showed that only between 20 and 30% of the participants use sex toys and the preferred ones seem to be plugs. The fantasies that most stimulate and excite our participants are having sex in public, having sex with more than one person at the same time, blindfolded sex, being tied up and observing a naked person.

As concerning anal sex, about 80–90% of the respondents answered that they did not practice it. A study by Habel et al., conducted in the general population of the United States between 2011–2015, reported that the prevalence of anal sex among heterosexual people is between 33–38%, which is slightly higher, compared to previous years [17]. It is possible that, in Italy, sexual freedom and the desire to experiment with new sexual practices might be overshadowed by a common feeling of shame. However, we have to consider the high rate of heterosexual respondents in our study.

As for pornography, our results are in line with the study of Herbenick et al. [5], with 80% of our participants watching porn at home, alone or from their smartphones.

Previously, we have demonstrated a positive association between porn addiction and erectile function, suggesting that a normal balancing between sexual activity, masturbation and pornography [18].

Furthermore, the impact of pornography on sexual behaviour is extremely important, expecially during previous COVID-19 pandemic.

In fact, different studies demonstrated an increased interest in pornography and coronavirus-themed pornography after the outbreak of COVID-19 in both eastern and western countries [19,20].

All these data reflect the development of faster internet connections and the pervasive distribution of smartphones, that have somehow partially replaced the use of larger computers and devices, making pornography even more easily and discreetly accessible from everywhere, at any time, and its importance in the context of sexual behaviour.

The most worrying data that arises from our survey is that only 66% of the respondents use condoms and only 37% use them regularly, when in a stable relationship. Since in our cohort the rate of occasional partners (10.5%) and polyamorous relationship (1.2%) were low, we believe is fundamental that new and more incisive awareness campaigns should be carried out, in order to avoid the spread of sexually transmitted diseases.

Strenght of the current study are represented by the inclusion of a large number of participants and to have investigated different aspect of sexuality and sexual behaviour. Our results could be useful for further researches in the field and to have a photography of sexual behaviuour of a young Italian population

Limitations of our study include the lack of investigation of older people, due to the use of Internet as source of enrolment, and the use of not standarized questionnaires. Futhermore, sexual habits and behaviours of may be different after COVID-19 pandemic and they should be taken into account by future researches. Finally, our cohort was young and it may be not representative of the general population.

5. Conclusions

Our survey was born with the aim of gaining a representative picture of sexuality in Italy before COVID-19 pandemic. There is still much to be done in order to increase people's awareness of sexual pleasure and get to the point of feeling free to express sexual desires to a partner, without fearing to be judged. But even more important, it is necessary to increase awareness campaigns for the prevention of sexually transmitted diseases, especially among young people, who are more at risk, since they have fewer stable relationships and therefore often relate with different sexual partners. Moreover, our results can have several practical implications for the areas of sex education, sexual health and to counteract sexual dysfunction during COVID-19 pandemic. Given the current trends of sexual habits, it is necessary to safeguard the health of young people and support them by increasing their sexual well-being.

Supplementary Materials: The following are available online at https://www.mdpi.com/article/10.3390/jcm10194327/s1, Table S1. Sexual pleasure patterns in the all population, Table S2: Contraception use in the all cohort, Table S3: Sexual intercourse patterns, Table S4: Dating app use in the all cohort.

Author Contributions: Conceptualization, A.C., D.G.; methodology, G.I.R.; validation, M.D.M., G.I.R., G.P., C.T., D.G., G.C., B.G., G.M., L.M., A.C.; formal analysis, G.I.R., M.D.M.; investigation, D.G.; resources, D.G.; data curation, A.C.; writing—original draft preparation, G.I.R.; writing—review and editing, G.I.R., M.D.M.; visualization, M.D.M., G.I.R., G.P., C.T., D.G., G.C., B.G., G.M., L.M., A.C.; supervision, M.D.M., G.I.R., G.P., C.T., D.G., G.C., B.G., G.M., L.M., A.C. All authors have read and agreed to the published version of the manuscript.

Funding: This research received no external funding.

Institutional Review Board Statement: This research has been conducted in the public arena using only publicly available or accessible records without contact with the individual/s and it does not require ethics reviews according to Tri-Council Policy Statement: Ethical Conduct for Research Involving Humans Articles 2.2 to 2.4.

Informed Consent Statement: Written informed consent has been obtained from the participants to publish this paper.

Data Availability Statement: Data are available upon request.

Conflicts of Interest: The authors declare no conflict of interest.

References

1. Döring, N.; Poeschl, S. Experiences with Diverse Sex Toys among German Heterosexual Adults: Findings from a National Online Survey. *J. Sex Res.* **2019**, *57*, 885–896. [CrossRef] [PubMed]
2. Towe, M.; El-Khatib, F.; Osman, M.; Huynh, L.; Carrion, R.; Ward, S.; Reisman, Y.; Serefoglu, E.C.; Pastuszak, A.; Yafi, F.A. "Doc, if it were you, what would you do?": A survey of Men's Health specialists' personal preferences regarding treatment modalities. *Int. J. Impot. Res.* **2020**, *33*, 303–310. [CrossRef] [PubMed]
3. Bulot, C.; Leurent, B.; Collier, F. Pornography sexual behaviour and risk behaviour at university. *Sexologies* **2015**, *24*, e78–e83. [CrossRef]
4. Lewis, R.; Tanton, C.; Mercer, C.H.; Mitchell, K.R.; Palmer, M.; Macdowall, W.; Wellings, K. Heterosexual Practices among Young People in Britain: Evidence from Three National Surveys of Sexual Attitudes and Lifestyles. *J. Adolesc. Health* **2017**, *61*, 694–702. [CrossRef] [PubMed]
5. Herbenick, D.; Fu, T.-C.; Wright, P.; Paul, B.; Gradus, R.; Bauer, J.; Jones, R. Diverse Sexual Behaviors and Pornography Use: Findings from a Nationally Representative Probability Survey of Americans Aged 18 to 60 Years. *J. Sex. Med.* **2020**, *17*, 623–633. [CrossRef] [PubMed]
6. Miranda, E.P.; Taniguchi, H.; Cao, D.L.; Hald, G.M.; Jannini, E.A.; Mulhall, J.P. Application of Sex Aids in Men with Sexual Dysfunction: A Review. *J. Sex. Med.* **2019**, *16*, 767–780. [CrossRef] [PubMed]
7. Tortora, C.; D'Urso, G.; Nimbi, F.M.; Pace, U.; Marchetti, D.; Fontanesi, L. Sexual Fantasies and Stereotypical Gender Roles: The Influence of Sexual Orientation, Gender and Social Pressure in a Sample of Italian Young-Adults. *Front. Psychol.* **2019**, *10*, 2864. [CrossRef]
8. Rosina, A.; Fraboni, R. Is marriage losing its centrality in Italy? *Demogr. Res.* **2004**, *11*, 149–172. [CrossRef]
9. Luciano, M.; Sampogna, G.; del Vecchio, V.; Giacco, D.; Mulè, A.; de Rosa, C.; Fiorillo, A.; Maj, M. The family in Italy: Cultural changes and implications for treatment. *Int. Rev. Psychiatry* **2012**, *24*, 149–156. [CrossRef] [PubMed]
10. Caltabiano, M.; Rosina, A.; Dalla-Zuanna, G. Interdependence between sexual debut and church attendance in Italy. *Demogr. Res.* **2006**, *14*, 453–484. [CrossRef]
11. Ross, M.W.; Månsson, S.-A.; Daneback, K. Prevalence, Severity, and Correlates of Problematic Sexual Internet Use in Swedish Men and Women. *Arch. Sex. Behav.* **2012**, *41*, 459–466. [CrossRef] [PubMed]
12. Dunkels, E.; Franberg, G.-M.; Hallgren, C.; Daneback, K.; Löfberg, C. Youth, Sexuality and the Internet. In *Youth Culture and Net Culture*; IGI Global: Hershey, PA, USA, 2011; pp. 190–206. [CrossRef]
13. Mollaioli, D.; Sansone, A.; Ciocca, G.; Limoncin, E.; Colonnello, E.; Di Lorenzo, G.; Jannini, E.A. Benefits of Sexual Activity on Psychological, Relational, and Sexual Health During the COVID-19 Breakout. *J. Sex. Med.* **2021**, *18*, 35–49. [CrossRef] [PubMed]
14. Eysenbach, G. Improving the quality of Web surveys: The Checklist for Reporting Results of Internet E-Surveys (CHERRIES). *J. Med. Internet Res.* **2004**, *6*, e34. [CrossRef] [PubMed]
15. Salonia, A.; Bettocchi, C.; Boeri, L.; Capogrosso, P.; Carvalho, J.; Cilesiz, N.C.; Cocci, A.; Corona, G.; Dimitropoulos, K.; Gül, M.; et al. European Association of Urology Guidelines on Sexual and Reproductive Health—2021 Update: Male Sexual Dysfunction. *Eur. Urol.* **2021**, *80*, 333–357. [CrossRef] [PubMed]
16. Dewitte, M.; Reisman, Y. Clinical use and implications of sexual devices and sexually explicit media. *Nat. Rev. Urol.* **2021**, *18*, 359–377. [CrossRef] [PubMed]
17. Habel, M.A.; Leichliter, J.S.; Dittus, P.J.; Spicknall, I.H.; Aral, S.O. Heterosexual anal and oral sex in adolescents and adults in the United States, 2011–2015. *Sex. Transm. Dis.* **2018**, *45*, 775–782. [CrossRef] [PubMed]

18. Russo, G.I.; Campisi, D.; Di Mauro, M.; Falcone, M.; Cocci, A.; Cito, G.; Verze, P.; Capogrosso, P.; Fode, M.; Cacciamani, G.; et al. The impact of asexual trait and porn addiction in a young men healthy cohort. *Andrologia* **2021**, *53*, e14142. [CrossRef]
19. Zattoni, F.; Gül, M.; Soligo, M.; Morlacco, A.; Motterle, G.; Collavino, J.; Barneschi, A.C.; Moschini, M.; Moro, F.D. The impact of COVID-19 pandemic on pornography habits: A global analysis of Google Trends. *Int. J. Impot. Res.* **2020**. [CrossRef]
20. Ibarra, F.P.; Mehrad, M.; Mauro, M.D.; Godoy, M.F.P.; Cruz, E.G.; Nilforoushzadeh, M.A.; Russo, G.I. Impact of the COVID-19 pandemic on the sexual behavior of the population. The vision of the east and the west. *Int. Braz. J. Urol.* **2020**, *46*, 104–112. [CrossRef]

Journal of
Clinical Medicine

Article

The Impact of Lockdown on Couples' Sex Lives

Elisabetta Costantini [1], Francesco Trama [1,*], Donata Villari [2], Serena Maruccia [3], Vincenzo Li Marzi [2],
Franca Natale [4], Matteo Balzarro [5], Vito Mancini [6], Raffaele Balsamo [7], Francesco Marson [8], Marianna Bevacqua [9],
Antonio Luigi Pastore [10], Enrico Ammirati [11], Marilena Gubbiotti [12], Maria Teresa Filocamo [13],
Gaetano De Rienzo [14], Enrico Finazzi Agrò [15], Pietro Spatafora [2], Claudio Bisegna [2], Luca Gemma [2],
Alessandro Giammò [11], Alessandro Zucchi [16], Stefano Brancorsini [17], Gennaro Ruggiero [18] and Ester Illiano [1]

[1] Department of Surgical and Biomedical Science, Andrological and Urogynecological Clinic,
Santa Maria Terni Hospital, University of Perugia, 05100 Terni, Italy; elisabetta.costantini@unipg.it (E.C.);
ester.illiano@inwind.it (E.I.)

[2] Urology Clinic, Careggi Hospital, University of Florence, 50100 Florence, Italy; donata.villari@unifi.it (D.V.);
limarzi2012@gmail.com (V.L.M.); pietro.spatafora@hotmail.it (P.S.); claudio.bisegna@gmail.com (C.B.);
gemmaluca.dr@gmail.com (L.G.)

[3] Istituti Clinici Zucchi, 20861 Brughiero, Italy; serena.maruccia@gmail.com

[4] Urogynecologic Unit, S. Carlo Hospital, 00100 Rome, Italy; francanatale3@gmail.com

[5] Urology Clinic, University of Verona, 37100 Verona, Italy; matteo.balzarro@aovr.veneto.it

[6] Urology Clinic, University of Foggia, 71100 Foggia, Italy; mancini.uro@gmail.com

[7] Urology Clinic, Monaldi Hospital, 80125 Naples, Italy; r.bals@virgilio.it

[8] Urology Clinic, Chieri Carmagnola and Moncalieri Hospital, 10121 Turin, Italy;
francescomarson84@gmail.com

[9] Urology Clinic, Bianchi-Malacrino-Morelli Hospital, 89121 Reggio Calabria, Italy;
mariannachiarabevacqua@gmail.com

[10] Urology Clinic, ICOT, Department of Medical Surgical Sciences and Biotechnologies,
Sapienza University of Rome, 04100 Latina, Italy; antopast@hotmail.com

[11] Urology Clinic, Città della Salute e della Scienza, 10121 Turin, Italy; ammirati.enrico@gmail.com (E.A.);
giammo.alessandro@gmail.com (A.G.)

[12] Urology Clinic, San Donato Arezzo Hospital, 52100 Arezzo, Italy; marilena.gubbiotti@gmail.com

[13] Urology Clinic, ASL CN1, 12038 Savigliano, Italy; mt.fil@libero.it

[14] Urology Clinic, University of Bari, 70121 Bari, Italy; gaetanoderienzo@gmail.com

[15] Urology Clinic, Torvergata University, 00100 Rome, Italy; finazzi.agro@med.uniroma2.it

[16] Urology Clinic, University of Pisa, 56121 Pisa, Italy; zucchi.urologia@gmail.com

[17] Department of Experimental Medicine, University of Perugia, 05100 Terni, Italy; stefano.brancorsini@unipg.it

[18] Department of Psychology, Laboratory of Cognitive Science and Immersive Virtual Reality,
Univerity of Campania, Luigi Vanvitelli, 81100 Caserta, Italy; gennaro.ruggiero@unicampania.it

* Correspondence: francescotrama@gmail.com

Citation: Costantini, E.; Trama, F.;
Villari, D.; Maruccia, S.; Li Marzi, V.;
Natale, F.; Balzarro, M.; Mancini, V.;
Balsamo, R.; Marson, F.; et al. The
Impact of Lockdown on Couples' Sex
Lives. *J. Clin. Med.* **2021**, *10*, 1414.
https://doi.org/10.3390/jcm10071414

Academic Editors: Giorgio I. Russo
and Du Geon Moon

Received: 28 February 2021
Accepted: 23 March 2021
Published: 1 April 2021

Abstract: Background: the aim of this study was to perform an Italian telematics survey analysis on the changes in couples' sex lives during the coronavirus disease 2019 (COVID-19) lockdown. Methods: a multicenter cross sectional study was conducted on people sexually active and in stable relationships for at least 6 months. To evaluate male and female sexual dysfunctions, we used the international index of erectile function (IIEF-15) and the female sexual function index (FSFI), respectively; marital quality and stability were evaluated by the marital adjustment test (items 10–15); to evaluate the severity of anxiety symptoms, we used the Hamilton Anxiety Rating Scale. The effects of the quarantine on couples' relationships was assessed with questions created in-house. Results: we included 2149 participants. The sex lives improved for 49% of participants, particularly those in cohabitation; for 29% it deteriorated, while for 22% of participants it did not change. Women who responded that their sex lives deteriorated had no sexual dysfunction, but they had anxiety, tension, fear, and insomnia. Contrarily, men who reported deteriorating sex lives had erectile dysfunctions and orgasmic disorders. In both genders, being unemployed or smart working, or having sons were risk factors for worsening the couples' sex lives. Conclusion: this study should encourage evaluation of the long-term effects of COVID-19 on the sex lives of couples.

Keywords: Covid-19; lockdown; male sexual dysfunction; female sexual dysfunction

1. Introduction

On 21 February, 2020, in Italy, the first cases of severe acute respiratory syndrome coronavirus 2 (SARS-CoV-2)—the coronavirus responsible for coronavirus disease 2019 (COVID-19)—were documented. The number of cases quickly increased, leading to a pandemic. On 10 March, 2021, a total of 487,074 cases and 100,811 deaths were reported in Italy [1].

On 5 March, 2020, a national lockdown was declared (Phase 1). For 50 days, this lockdown affected all national production sectors and health services; non-urgent ambulatorial [2] and surgical activities [3] were suspended in all Italian hospitals.

The restrictions prevented families, friends, and sometimes non-cohabiting couples from physically meeting. On 4 May, 2020 "Phase 2" began, which allowed people to meet family members and relatives living in the same city, but other restrictions were unchanged. The lockdown has impacted the entire population; people of all age groups have changed their habits, which has led to increased uncertainty about the future, especially in regards to (often irreversible) changes, such as job loss.

In Italy, according to the National Institute of Statistics (ISTAT) [4], after the substantial stagnation of the first two months in 2020 (-0.1% in January and $+0.1\%$ in February), the onset of the pandemic hit the work market, causing a reduction of 124,000 employees (-0.5%) in March, more than double that number in April (-274 thousand, -1.2%), and a continuation in May (-84 thousand, -0.4%). The job market and financial insecurity were related to symptoms of depression and anxiety [5,6]. Repercussions of the lockdown on the Italian population's sex lives are less known. In Italy, crises that have changed peoples' habits (e.g., the L'Aquila earthquake) have been studied. It was shown that crises can alienate loved ones; moreover, the loss of a home can alter daily habits and couples' sex lives. After the earthquake, there was a high rate of sexual dysfunction-related symptoms in young adults, particularly subjects experiencing post-traumatic stress disorder [7]. The lockdown has also likely led to changes in the sex lives of Italians.

Overall, evidence of the impacts of external stressful events on couples' sex lives is still being debated. Although a few studies have addressed the issue of sexual behavior during the pandemic [8–12], to our knowledge, there is not much data [13] on Italian couples' sex lives during events such as the lockdown that also investigated sexual activity and functioning. Since intimate relationships can be a reflection of the "goodness" of couples' psychological and physical states, we investigated if (and to what extent) uncertainty and perceived danger could, albeit temporarily, cause changes in the sex lives of Italian couples. Studying these factors could allow us to better understand the effects of social deprivation and of perceived/actual danger, as well as how couples are able to compensate for a long-term lack of basic psychological needs.

We hypothesize that the lockdown influenced the couples' sex lives. The aim of this study was to perform a telematics survey analysis of the changes in the sexual behavior of adult men and women in stable relationships during Phase 1 of the lockdown.

2. Methods

This was an Italian, multicenter cross-sectional study, conducted from 15 urological centers. It was approved by the local ethics committee, and participants signed online informed consent documents.

2.1. Participants

The study was conducted from 4 May 2020 (50 days after the start of the lockdown) to 18 May 2020. Inclusion criteria were: subjects sexually active, in stable relationships for at least 6 months, both sexes, and of any age. Exclusion criteria were: subjects who were COVID-19 positive, single, or sexually inactive.

2.2. Procedure

The research was performed with an online survey. A questionnaire was created in Italian via Google Forms (Supplementary Material 1). The questionnaire link was forwarded to all investigator associates. Respondents were recruited through convenience sampling and were asked to forward (or post) the links among their contact groups in all social networks (i.e., Facebook) or by free communication apps (i.e., WhatsApp).

Clicking on the questionnaire link caused the consent form and the instructions to appear on the screen. The questionnaire became accessible after participants accepted the terms and conditions of the study. Data cleaning was performed by one of the investigators and was cross-checked by a second investigator.

2.3. Measures

For each participant, demographic data were obtained, including age, gender, weight, and height, as freeform questions; and sexual orientation, current region of residence, and occupation as multiple choice questions. Other questions were: "Do you have children?" and "Do you live with your partner?" with dichotomous answers; and "How many years have you been in a relationship with your partner?".

In regards to sexual functioning, women had to fill in the female sexual function index (FSFI-19 items) [14], and the international index of erectile function (IIEF-15 items) [15] was used to evaluate female and male sexual function in the 4 weeks before the survey, during Phase 1. Married participants completed the marital adjustment test (MAT), items 10 to 15 [16], which evaluated marital quality and stability. The items chosen investigated couple complicity and the presence of common interests

All participants completed the Hamilton Anxiety Rating Scale (HAM) [17], which measures the severity of anxiety symptoms. The effects of the COVID-19 pandemic and quarantine on couples' relationships were assessed with questions created in-house (Table 1). We defined "improvement of sexual life" with the answer "Much; very much" to Item 4 ("Do you think that your couple's sex life has improved during this period?") of COVID-19; while "worsening of sexual life" with the answer "Much; very much" to Item 3 ("Do you think that your sex life as a couple has deteriorated during this period?") of COVID-19.

Table 1. Questions concerning the coronavirus disease 2019 (COVID-19) pandemic and its effect on couples' relationships.

	No	Not Much	So and So	Much	Very Much
1. Do you feel safe at home?					
2. Do you feel safe outside the home?					
3. Do you think that your sex life as a couple has deteriorated during this period?					
4. Do you think that your sex life as a couple has improved during this period?					
5. Do you feel safe with your partner at home?					
6. Do you feel dissatisfied with your partner at home?					
7. Do you feel happy with your partner at home?					
8. Do you feel uncomfortable with your partner at home?					
9. How comfortable do you feel with your partner at home?					
10. How satisfied do you feel with your partner at home?					
11. Do you think that your couple problems have decreased during this period?					
12. Do you feel unhappy with your partner at home?					
13. Do you think that your couple problems have increased during this period?					
14. Do you feel more nervous towards your partner during this period?					
15. Do you feel more calm towards your partner during this period?					

2.4. Statistical Analysis

All answers were downloaded via Google Form and reported in a calculation file, and each answer was converted into a corresponding score on the basis of the questionnaire analyzed. The Mann–Whitney test was used to compare ordinal and non-normally distributed continuous variables; to compare normally distributed continuous variables we used T-Test. Categorical data were analyzed with the $\chi 2$ test with Yates correction or Fisher's exact test. Multivariate logistic regression models were fit for the prediction of risk factors (clinical and demographic data) for worsening female and male sex lives. Multivariate logistic regression models were fit, incorporating all variable analyses in bivariate analyses. Odds ratios (ORs) with 95% CIs were also calculated. Statistical analyses were performed in software (SPSS, Version 23.0; IBM Corp, Armonk, NY, USA). A two-sided p value < 0.05 was considered significant. We did not perform the correction for multiplicity. The internal validity of the COVID-19 questionnaire was evaluated by Cronbach alpha test.

3. Results

From 4 May to 18 May 2020, we enrolled 2150 participants. One participant was excluded because the questionnaire had been filled out incorrectly. The analysis was performed on 2149 participants. Table 2 shows the demographic data of the study's total population and both genders. The sample size for homosexuals and bisexuals was too small to perform statistical analysis; therefore, the analysis was performed on the entire population regardless of sexual orientation. The COVID-19 questionnaire showed a high level of inter-item reliability and Cronbach's alpha (0.76).

Table 2. The demographic data of the general population and female and male participants.

Data	Total Population *n* (2149)	Female *n* (1112)	Male *n* (1037)
Age (mean $\pm$ SD)	43.07 $\pm$ 12.5	43 $\pm$ 12.5	43.2 $\pm$ 12.4
BMI (median, range)	24.16 (18.90–44.2)	24.14 (18.90–43.1)	24.21 (19.90–44.2)
Sexual Orientation Heterosexual *n* (%) Homosexual *n* (%) Bisexual *n* (%)	2035 (94) 91 (4) 23 (10)	1075 (96.6) 18 (16.2) 19 (17)	960 (92.5) 73 (7) 4 (0.4)
Son *n* (%)	1253 (58)	657 (59)	596 (57.4)
Residences North *n* (%) Central *n* (%) South and Islands *n* (%)	665 (31) 773 (36) 711 (33)	359 (32.2) 417 (37.5) 336 (30.3)	306 (29) 356 (34.3) 375 (36.7)
Education Primary school *n* (%) Secondary school *n* (%) High school *n* (%) Graduate school *n* (%)	7 (0.3) 132 (6) 712 (33) 1298 (60)	2 (0.1) 78 (7) 352 (31.7) 680 (61.2)	5 (0.4) 54 (5.2) 300 (28.9) 678 (65.3)
Occupation Student *n* (%) Retired *n* (%) Unemployed *n* (%) Working at the usual workplace *n* (%) Smart working *n* (%)	105 (4) 112 (5.2) 181 (8.4) 735 (3.4) 1016 (4.7)	46 (4) 50 (4.5) 127 (11.4) 441 (39.6) 448 (40.3)	59 (5.7) 62 (5.9) 54 (5.3) 294 (28.4) 568 (54.7)
Cohabitants *n* (%)	1667 (77.5)	895 (80)	772 (74.4)
Married	1238 (5.7)		
Years of stable relationships $\leq$1 years *n* (%) Da 1 a 3 years *n* (%) Da 3 a 5 years *n* (%) $\geq$5 years *n* (%)	171 (7.9) 282 (13) 204 (9.4) 1492 (69.4)	162 (14.5) 146 (1.43) 130 (1.16) 774 (69.6)	109 (10.5) 136 (13.2) 74 (7.1) 718 (69.2)
Questionnaire IIEF (mean $\pm$ SD) FSFI (median, range) MAT (mean $\pm$ SD) HAM (mean $\pm$ SD)	38.9 $\pm$ 28.2 28.5 (2–35.6) 50.3 $\pm$ 2.6 5.1 $\pm$ 1.3	45.9 $\pm$ 12.2 28.5 (2–35.6) 47.3 $\pm$ 6.2 6.3 $\pm$ 2.5	34.2 $\pm$ 13.5 28.5 (2–35.6) 49.7 $\pm$ 3.5 5.3 $\pm$ 3.8

IIEF: International index of erectile function; FSFI: female sexual function index MAT: Marital adjustment test;HAM: Hamilton Anxiety Rating Scale.

J. Clin. Med. **2021**, 10, 1414

A total of 49% replied "much or very much" to the question "Do you think that your couple's sex life has improved during this period?" (Item 4 COVID-19); 29% replied "much or very much" to the question "Do you think that your couple's sex life has deteriorated during this period?" (Item 3 COVID-19); and 22% did not report a change—subjects who answered that they neither had an improvement nor a worsening (Items 3 and 4, COVID-19).

Table 3 shows the demographic data of the general population who reported an improvement or worsening, or no changes of couples' sex lives.

Table 3. Demographic data on subjects who reported an improvement, worsening, or no changes in the couple's sex life.

Data	Total Improvement $n = 1049$	p Value	Total Worsening $n = 623$	p Value	Total No Change $n = 477$	p Value
Age						
>40 years (mean $\pm$ SD)	55.12 $\pm$ 3.2	0.2	57.25 $\pm$ 2.1	0.4	51.19 $\pm$ 3.8	0.4
<40 years (mean $\pm$ SD)	34.6 $\pm$ 4.1		37.4 $\pm$ 2.6		31.7 $\pm$ 2.4	
BMI	23.5 $\pm$ 1.2		25.4 $\pm$ 2.9		26.4 $\pm$ 3.6	
Gender						
Female n (%)	579 (55.2)	0.002	314 (50.4)	0.4	219 (45.9)	0.4
Male n (%)	470 (44.8)		309 (49.5)		258 (54.0)	
Years of stable relationships						
$\geq$5 year	764 (72.8)	<0.0001	399 (64.0)	<0.0001	148 (31.0)	0.0001
<5 years	285 (27.1)		224 (35.9)		329 (68.9)	
Married						
Yes n (%)	640 (61)	0.001	306 (49.1)	<0.0001	292 (61.2)	<0.0001
No n (%)	409 (38.9)		317 (50.8)		185 (38.7)	
Sexual Orientation						
Heterosexual n (%)	985 (93.8)	<0.0001	580 (93)	<0.0001	470 (98.5)	<0.0001
Homosexual n (%)	49 (4.6)		35 (5.6)		7 (1.4)	
Bisexual n (%)	15 (1.4)		8 (1.2)		0	
Educational						
Primary school n (%)	0	<0.0001	7 (0.12)	0.001	0	<0.0001
Secondary school n (%)	54 (5.14)		78 (12.5)		0 (0)	
High school n (%)	329 (31.3)		383 (61.4)		20 (4.1)	
Graduate school n (%)	985 (94)		313 (50.2)		457 (95.8)	
Residence						
North n (%)	313 (29.8)	0.007	288 (46.2)	<0.0001	64 (13.4)	0.01
South n (%)	324 (30.9)		218 (35.0)		231 (48.2)	
Centre n (%)	412 (39.3)		117 (28.4)		182 (38.1)	
Son						
Yes n (%)	423 (40.3)	0.112	334 (53.6)	0.003	30 (6.2)	<0.0001
No n (%)	626 (59.7)		289 (46.4)		447 (93.7)	
Occupation						
Student n (%)	50 (0.2)	0.001	80 (12.8)	<0.0001	0	<0.0001
Retired n (%)	85 (8.1)		10 (1.6)		17 (3.5)	
Unemployed n (%)	20 (2)		90 (14.4)		86 (18)	
Working at the usual workplace n (%)	650 (61.9)		20 (3.2)		65 (13.6)	
Smart working n (%)	244 (23.2)		423 (67.8)		309 (64.7)	
Questionnaire						
IIEF (mean $\pm$ SD)	37.4 $\pm$ 26.3		27 $\pm$ 28		32.6 $\pm$ 12	
FSFI (median, range)	28.5 (2–35.6)		28.5 (2–35.6)		28.5 (2.35–6)	
MAT (mean $\pm$ SD)	64.3 $\pm$ 2.1		42.7 $\pm$ 5.4		48.2 $\pm$ 2.6	
HAM (mean $\pm$ SD)	5.2 $\pm$ 3.5		9.7 $\pm$ 5.1		5.1 $\pm$ 2.3	

Participants who reported improved sex lives were mostly female ($p = 0.01$) or participants with lasting relationships ($p < 0.0001$), compared to those who reported deteriorating sex lives, or having no changes during the lockdown.

Subjects who reported worsening sex lives were unmarried ($p = 0.01$) or had sons ($p < 0.0001$) compared to other groups.

Male subjects ($p = 0.01$), participants with shorter relationships ($p < 0.0001$), or participants without sons ($p < 0.0001$) reported no changes compared to participants in others groups.

There was no statistical difference according to age.

3.1. The Improvement of Couples' Sex Lives

Improvements in couples' sex lives was reported more by female subjects than male ($p = 0.002$); improvements were also associated with the following variables: cohabiting with the partner (84%), being in a stable relationship for more than 5 years (72.8%), married, without sons (59.7%). Both men and women who reported an improvement in their sex lives (Item 4 COVID-19) had good relationships with their partners.

Among women, 97% of them had all (or some) interests in common with their partners (Item 2 MAT (Marital Adjustment Test)), and 65.7% liked to do the same activities as their partners in their free time (they gave the same answers to Items 4–5 in the MAT), and 70% would remarry the same person (Item 6 MAT).

Among men, 72% and 74% were happy (Item 7 COVID-19) and satisfied with their partners (Item 10 COVID-19), respectively; 83% had all or some interests in common with their partners (Item 2 MAT), and 66% liked to do the same activities as their partners in their free time (Items 4–5 MAT). A total of 80% would remarry the same person (Item 7 MAT).

None of the subjects, in either gender, had sexual dysfunctions. Men experienced good erectile and orgasmic functions, high sexual desire, and overall satisfaction (IIEF mean 33.5 ± 21.4) in the prior 4 weeks, while women had a median FSFI score of 28.5 (2–35.6). They lived mostly in central (39.3%) or south (30.9%) Italy; most attended graduate school (94%) and worked at the usual workplaces (61.9%).

3.2. Worsening of Couples' Sex Lives

Worsening of couples' sex lives was reported in both sexes, but without statistically significant differences between genders ($p = 0.4$) or between people under and over 40 years of age (47.4% vs. 48.6%, $p = 0.8$). Most of these subjects did not live with their partners during lockdown (73.4%). Among those who reported worsening sex lives (as a couple) and who lived with their partners (26.6%), 82% had sons, and 81.7% had stable relationships for more than 5 years. In 50.8% of the cases, they lived in north Italy; most attended high school (55.7%), and worked at home in smart working (68.7%).

Women who responded that "their couple's sex lives has deteriorated" (Item 3 COVID-19) had no sexual dysfunction (FSFI median score 28.5 (2–35.6)) (Table 4). However, they had higher anxiety (17.7% vs. 8.4%, $p < 0.0001$, item 1 HAM score), tension (21.6% versus 10.5%, $p < 0.0001$, Item 2 HAM score), fear (21.3% vs. 7.2% $p < 0.0001$, Item 3 HAM score), and insomnia (27.5% vs. 3.7%, $p < 0.0001$, Item 4 HAM score) than women who had replied "no, not much or so-so" to the question of whether their sex life had worsened. In addition, among cohabitants, women were also more likely to be dissatisfied with their partners (13.8% vs. 5.3%, $p < 0.0001$) and to feel nervous toward their partners during this period (24.6% vs. 5.1%, $p < 0.0001$) than women who had replied "no, not much or so-so" to the question on whether their sex life had worsened.

In contrast, men who reported worsened couples' sex lives had mild erectile dysfunctions, orgasmic dysfunctions, and low sexual satisfaction in the prior 4 weeks (Table 4). These results were conformed in the univariate and multivariate analysis (Table 5); in fact, erectile dysfunction, orgasmic dysfunction, and low intercourse satisfaction were risk factors for worsened couples' sex lives. Cohabitating men (57%) had mild erectile dysfunction (19 ± 10.5), pathological scores of desire (of 5.5 ± 2.1), and overall satisfaction of 5.5 ± 2.1. They felt more uncomfortable with their partners at home than those who did not report worsened sex lives (12.9% vs. 5.4%, $p < 0.0001$, item 8 COVID-19), and had mild anxiety symptoms (median score 17 (8–32)). However, only 11.3% of married people said that they would marry another person (Item 6 MAT Test), 19.5% experienced an increase in couples' problems (Item 13 COVID-19), and 7% were unhappy with their partners at home (Item 12 COVID-19). Among non-cohabitants the IIEF scores, except for sexual desire (Table 4) and moderate symptoms of anxiety (median score 29 (8–35)), were associated with a worsening of couples' sex lives.

Table 4. Demographic data on subjects who lived with their partners during the lockdown, and reported an improvement, worsening, or no changes in the couple's sex life.

Data	Total Improvement; $n = 883$	p Value	Total Worsening; $n = 393$	p Value	Total No Changes $n = 389$	p Value
Age >40 years <40 years	54.09 ± 3.1 36.4 ± 3.6	0.2	53.39±2.3 35.3±3.6	0.3	56.23 ± 3.4 34.2 ± 2.7	0.3
BMI	24.7± 2.9		26.3±3.6		27.4±3.1	
Gender Female *n* (%) Male *n* (%)	491 (55.6) 392 (44.4)	0.001	217 (55.2) 176 (44.8)	0.002	200 (51.4) 189 (48.5)	0.01
Years of stable relationships ≥5 year <5 years	721 (81.7) 162 (18.3)	<0.0001	321 (81.7) 721(18.3)	<0.0001	310 (79.6) 79 (20.3)	<0.0001
Married Yes *n* (%) No *n* (%)	628 (71.1) 255 (28.8)	<0.0001	263 (66.9) 130 (33.0)	<0.0001	256 (65.8) 133 (34.1)	<0.0001
Educational Primary school *n* (%) Secondary school *n* (%) High school *n* (%) Graduate school *n* (%)	0 53 (6) 300 (33.9) 530 (60.1)	<0.0001	4 (1.01) 70(17.8) 219(55.7) 100(25.4)	<0.0001	2 (0.5) 189 (48.5) 98 (25.1) 100 (25.7)	0.001
Residence North *n* (%) South *n* (%) Centre *n* (%)	229 (25.9) 254 (28.7) 400 (45.3)	0.005	200 (50.8) 100 (25.4) 3 (23.6)	0.001	75 (19.2) 130 (33.4) 184 (47.3)	<0.0001
Son Yes *n* (%) No *n* (%)	400 (45.3) 483 (54.7)	0.001	325 (82) 68 (17.3)	<0.0001	135 (34.7) 254 (65.3)	<0.0001
Occupation Student *n* (%) Retired *n* (%) Unemployed *n* (%) Working at the usual workplace *n* (%) Smart working *n* (%)	0 (0.2) 100 (11.3) 13 (1.4) 670 (75.8) 100 (11.3)	0.001	0 (12.8) 3 (0.7) 100 (25.4) 20 (5.1) 270(68.7)	0.001	125 (32.1) 112 (28.7) 0 100 (25.7) 25 (6.4)	0.01
Sexual Orientation Heterosexual *n* (%) Homosexual *n* (%) Bisexual *n* (%)	875 (93.8) 8 (0.9) 0	<0.0001	384 (97.7) 0 9 (2.3)	<0.0001	388(99.7) 1 (0.25) 0	<0.0001
Questionnaire IIEF (mean ± SD) FSFI (median, range) MAT (mean ± SD) HAM (mean ± SD)	33.5 ± 21.4 28.5 (2–35.6) 75.3 ± 3.5 5.1 ± 2.4		25 ± 12 28.5 (2–35.6) 41.7 ± 6.4 9.2 ± 4.3		34.8 ± 10.2 28.5 (2.35–6) 46.2 ± 2.4 5.1 ± 2.7	

Tables 6 and 7 showed the results of univariate analysis and multivariate logistic regression for risk factor assessment for worsening female and male sex lives, respectively. In both genders, being unemployed or smart working during this period, as well as living in north Italy, having sons, and a relationship for more than 5 years, were risk factors for worsened sex lives.

Table 5. Univariate analysis and multivariate logistic regression final model for female worsening sex lives vs. demographic and psychological data.

Female Worsening Sexual Life	Univariate Analysis		Logistic Regression	
	p Value	OR (95% CI)	*p* Value	OR (95% CI)
Age	0.63	0.56 (0.71–1.22)	0.98	0.98 (0.87–1.15)
BMI	0.77	0.66 (0.50–1.89)	0.51	0.66 (0.50–1.89)
Sexual Orientation				
Heterosexual	0.13	1.71 (0.74–3.94)	0.24	1.96 (0.62–6.15)
Homosexual	0.2	0.50 (0.14–1.75)	0.75	0.75 (0.13–4.19)
Bisexual	0.34	0.67 (0.22–2.04)	0.54	0.85 (0.19–1.98)
Son				
Yes	0.001	1.82 (1.63–2.09)	0.001	1.85 (1.68–2.21)
No	0.2	0.94 (0.84–1.57)	0.12	0.87 (0.75–1.10)
Residences				
North	0.002	1.50 (1.14–1.97)	0.004	1.63 (1.16–2.28)
Central	0.01	0.64 (0.48–0.84)	0.002	1.52 (1.14–2.37)
South and Islands	0.4	1.04 (0.78–1.38)	0.4	1.16 (0.81–1.65)
Education				
Primary school	0.34	1.12 (0.87–1.24)	0.49	1.22 (1.11–1.78)
Secondary school	0.45	1.20 (1.01–1.36)	0.67	1.32 (1.20–1.55)
High school	0.67	1.17 (1.08–1.45)	0.82	1.24 (1.14–1.37)
Graduate school	0.55	1.24 (1.15–1.57)	0.76	1.31 (1.20–1.68)
Occupation				
Student	0.06	1.67 (0.91–3.06)	0.08	1.57 (1.10–3.58)
Retired	0.06	0.27 (0.10–0.68)	0.5	0.8 (0.4–1.67)
Unemployed	0.02	1.62 (1.19–2.35)	0.04	1.45 (1.20–3.25)
Working at the usual workplace	0.39	1.04 (0.82–1.36)	0.4	0.75 (0.31–1.52)
Smart working	0.001	1.27 (1.10–1.68)	0.01	1.32 (1.15–1.76)
Married				
Yes	0.04	1.28 (0.98–1.67)	0.83	1.35 (1.10–1.57)
No	0.04	0.78 (0.59–1.01)	0.78	0.85 (0.64–1.21)
Years of stable relationships				
<5 years	0.04	0.77 (0.58–1.02)	0.7	0.94 (0.74–3.45)
≥5 years	0.04	1.29 (0.98–1.71)	0.02	1.49 (1.13–4.58)
Psychological data				
Anxiety	<0.0001	2.36 (1.60–3.48)	0.03	1.28 (0.78–2.09)
Tension	<0.0001	2.34 (1.64–3.34)	0.13	3.27 (2.51–5.34)
Fear	<0.0001	2.33 (1.63–3.30)	0.001	2.57 (1.78–4.16)
Insomnia	<0.0001	2.34 (1.64–3.34)	0.04	1.41 (0.94–2.13)

Table 6. Univariate analysis and multivariate logistic regression final model for male worsened sex lives vs. demographic and psychological data.

Male Worsening Sexual Life	Univariate Analysis		Logistic Regression	
	p Value	OR (95% CI)	*p* Value	OR (95% CI)
Age	0.74	0.67 (0.41–1.34)	0.89	0.78 (0.57–1.29)
BMI	0.59	0.58 (0.34–1.91)	0.66	0.61 (0.34–1.97)

Table 6. *Cont.*

Male Worsening Sexual Life	Univariate Analysis		Logistic Regression	
	p Value	OR (95% CI)	*p* Value	OR (95% CI)
Sexual Orientation				
Heterosexual	0.25	1.39 (0.45–4.26)	0.35	1.57 (0.74–4.23)
Homosexual	0.45	0.72 (0.38–1.98)	0.89	0.87 (0.34–5.23)
Bisexual	0.55	0.45 (0.27–2.59)	0.74	0.82 (0.32–2.65)
Son				
Yes	0.8	1.94 (1.54–2.23)	0.24	1.86 1.78–2.98)
No	0.8	0.92 (0.89–1.64)	0.24	0.97 (0.87–2.10)
Residences				
North	0	2.57 (1.25–2.58)	0.002	1.81 (1.24–2.59)
Central	0.7	0.86 (0.57–1.89)	0.003	1.48 (1.17–2.42)
South and Islands		1.44 (0.82–1.76)	0.35	1.78 (0.43–2.87)
Education				
Primary school	0.54	1.46 (0.91–3.24)	0.74	1.51 (1.23–2.14)
Secondary school	0.61	1.29 (1.17–1.47)	0.89	1.52 (1.20–2.18)
High school	0.85	1.28 (1.04–1.59)	0.94	2.14 (1.87–2.69)
Graduate school	0.45	1.78 (1.35–2.25)	0.87	1.54 (1.36–1.92)
Occupation				
Student	0.09	1.85 (1.25–3.78)	0.06	1.94 (1.20–4.58)
Retired	0.05	0.74 (0.21–0.84)	0.3	1.45 (1.15–1.82)
Unemployed	0.04	1.83 (1.49–2.75)	0.03	1.95 (1.54–3.25)
Working at the usual workplace	0.7	1.30 (0.47–2.41)	0.6	0.84 (0.45–1.78)
Smart working	0.01	1.57 (1.65–2.69)	0.02	3.24 (1.55–3.98)
Married				
Yes	0.02	1.36 (0.76–2.47)	0.74	1.98 (1.22–2.36)
No	0.02	0.95 (0.48–2.58)	0.54	0.93 (0.74–2.14)
Years of stable relationships				
<5years	0.03	0.48 (0.10–1.58)	0.4	1.24 (0.94–2.69)
≥5 years	0.03	1.78 (0.68–2.36)	0.01	2.36 (1.25–3.45)
Psychological data				
Anxiety	<0.0001	1.56 (0.45–2.87)	0.01	1.78 (0.92–3.45)
Tension	<0.0001	2.71 (1.36–3.56)	0.78	4.13 (2.63–6.21)
Fear	<0.0001	2.45 (1.79–4.56)	0.02	2.96 (1.61–4.57)
Insomnia	<0.0001	2.36 (1.87–4.51)	0.07	2.57 (1.45–3.68)

Table 7. Univariate analysis and multivariate logistic regression final model for female cohabitants worsening sex lives vs. demographic and psychological data.

Female Worsening Sexual Life	Univariate Analysis		Logistic Regression	
	p Value	OR (95% CI)	*p* Value	OR (95% CI)
Age	0.005	0.62 (0.45–0.86)	0.02	0.92 (0.65–1.23)
BMI	0.54	1.37 (1.01–1.79)	0.63	1.67 (1.35–1.86)
Sexual Orientation				
Heterosexual	0.25	1.85 (0.69–4.25)	0.34	1.93 (0.84–5.21)
Homosexual	0.36	0.75 (0.23–1.89)	0.87	0.84 (0.65–4.63)
Bisexual	0.65	0.94 (0.34–2.37)	0.78	0.61 (0.32–1.85)
Son				
Yes	0.51	1.29 (0.87–1.88)	0.26	1.74 (0.45–1.36)
No	0.32	0.66 (0.84–1.75)	0.26	0.82 (0.47–1.68)

Table 7. *Cont.*

Female Worsening Sexual Life	Univariate Analysis		Logistic Regression	
	p Value	OR (95% CI)	p Value	OR (95% CI)
Residences				
North	0.001	1.74 (1.35–2.35)	0.002	1.78 (1.45–2.80)
Central	0.03	0.53 (0.10–0.95)	0.001	1.92 (1.32–2.88)
South and Islands	0.5	1.26 (0.95–1.47)	0.6	1.45 (0.96–1.84)
Education				
Primary school	0.65	1.45 (0.95–1.63)	0.88	1.36 (1.14–1.95)
Secondary school	0.57	1.38 (1.10–1.78)	0.69	1.57 (1.32–1.86)
High school	0.82	1.26 (1.04–1.83)	0.71	1.49 (1.26–1.55)
Graduate school	0.63	1.87 (1.34–1.93)	0.86	1.61 (1.47–1.73)
Occupation				
Student	0.08	1.54 (0.84–3.54)	0.5	1.89 (1.30–3.84)
Retired	0.07	0.59 (0.16–0.84)	0.7	0.9 (0.12–1.75)
Unemployed	0.01	1.87 (1.36–2.73)	0.02	1.59 (1.25–3.59)
Working at the usual workplace	0.5	1.65 (0.69–1.95)	0.8	0.96 (0.47–1.73)
Smart working	0.002	1.43 (1.05–1.87)	0.01	1.63 (1.11–1.94)
Married				
Yes	0.03	0.94 (0.45–1.36)	0.91	0.75 (0.35–1.67)
No	0.03	1.45 (0.76–1.62)	0.84	1.61 (1.32–1.99)
Years of stable relationships				
<5 years	0.01	0.84 (0.41–1.57)	0.5	0.84 (0.54–3.75)
≥5 years	0.01	1.65 (0.86–1.92)	0.03	1.81 (1.35–4.98)
Psychological data				
Anxiety	<0.0001	2. 62 (1.72–4.11)	0.04	1.65 (0.52–2.79)
Tension	<0.0001	2.75 (1.75–4.80)	0.25	3.58 (2.47–5.88)
Fear	<0.014	1.01 (0.99–1.03)	0.02	2.67 (1.45–4.53)
Insomnia	<0.0001	2.26 (1.48–3.45)	0.01	1.78 (0.68–2.90)

3.3. Risk Factors of Worsening of Couples' Sex Lives

In the univariate, being married was a risk factor for both women and men, but the result was not confirmed at the multivariate. Anxiety, fear, and insomnia that developed during this period appeared to be risk factors for worsened sex lives in both genders.

The same risk factors were obtained in the univariate and multivariate logistic regression for risk factor assessment for the worsened sex lives of female and male cohabitants.

A total of 76% of the population replied, "no or not much" to the question "Do you feel safe outside your home?" (Item 2 COVID-19). According to the HAM scores, participants feeling insecure while away from home had higher anxiety (88.2% vs. 11.5%, $p = 0.002$), tension, fatigue, alarm responses, crying, trembling, restlessness, inability to relax (90.7% vs. 9.3%, $p = 0.04$), fear (99.6% vs. 0.4% $p < 0.0001$), and insomnia (86.9% vs. 13.1%, $p = 0.04$) than participants who replied "much or very much."

A total of 90% of the respondents replied that "they felt very or very safe inside the home" (Item 1 COVID-19), and 84.5% that "they felt safe at home with their partner" (Item 5 COVID-19).

4. Discussion

The combination of isolation and risk of contagion has provoked a negative cumulative effect in terms of psychological and socio-cognitive resilience. Isolation is a slow stress factor because it prevents social relationships that, in turn, help people regulate their emotions, cope with stressful events, and strengthen their resilience during difficulties [18,19]. Consistently, in the current study, as in an another Chinese cross-sectional survey [10], the majority of participants (especially those who reported feeling insecure outside the home) reported high levels of anxiety, fear, agitation, feelings of restlessness, and insomnia.

Our study showed that the COVID-19 pandemic has also influenced Italian couples' sex lives. Simultaneously, the "lockdown" led to an improvement in couples' sex lives in 49% of participants, particularly cohabitants, whereas 29% reported a worsened sex life for different reasons between men and women, and 22% reported no change.

During the lockdown, despite the impossibility of meeting friends and relatives, and maintaining stability, in many cases, a rapprochement occurred among cohabiting couples. Most people reported that they were satisfied and happy with their partners at home. The improvement was reported primarily in participants who had been in stable relationships for more than 5 years, probably because the increased time spent together favored the rediscovery of a feeling that the couple might have lost in their life routines. Spending entire days at home can stimulate and facilitate common interests between partners—the sharing of hobbies or daily practices that normally could not be shared because of a lack of time.

Participants over the age of 40 improved more than those younger than 40, probably because most of the younger participants did not live with their partners during this period.

In our study, in both genders, participants who reported an improvement in their sex lives did not have sexual dysfunction. It allowed them to obtain qualitatively valid sexual activity, which strengthened the couples, as such. Moreover, another Italian survey conducted on participants younger than those in the present study, showed improvements in the frequency of sexual intercourse and sexual desire in both genders [13]. Villani showed that the increase in frequency of sexual intercourse also caused an increase in spontaneous pregnancy during the lockdown [20].

The worsening of couples' sex lives was reported in both genders, without a statistically significant difference between them (women 50.4% versus men 49.6%). In both genders, risk factors included being unemployed or smart working during this period, as well as living in north and central Italy, having sons, and a relationship for more than 5 years. The absence of children also likely allowed more time for couples. The partners were thus able to act on their sexual interests at any time. In everyday life, children attend school and, at times, are cared for by their grandparents after-school, and boys often engage in activities that allow parents to keep their space. However, the pandemic has changed the status quo, forcing couples to find fleeting moments of intimacy, which may not always be possible. During the pandemic period, couples with children were engaged in childcare and distance learning, and this negatively influenced the time available to devote to their partners and married lives.

Our study did not investigate the presence, during the lockdown, of other family members besides children; of course, this could also affect a couple's sex life. However, we assumed that the decree allowed them to live only between cohabitants, and consequently, any other family members were present even before the lockdown; therefore, there should be no such influence to be analyzed.

Women who reporting a worsening of the couple's sex life had no pathological median FSFI score, but they had emotional difficulties. Presumably, the worsening of women's sex lives can be attributed to several factors, including the lockdown, psychological and sociocultural factors, and interpersonal well-being [21–23]. The pandemic, similar to other catastrophic situations (such as earthquakes, hurricanes or wars), could in fact cause anxiety and depression, thus, decrease the frequency of sexual intercourse [24], sexual desire [25], libido, orgasm, and vaginal lubrication [26,27].

In men, worsened sex lives were predominately due to sexual dysfunction; however, a low percentage of men were unhappy at home with their partners, and would marry other people. In this frame, presumably, we cannot exclude that, along with the lockdown effect, the link between sexual dysfunction and psychological well-being [28], cognitive attributions, depressive–anxiety state, partner reactions, automatic failure/disengagement thoughts, and ineffective coping styles may have played a critical role [29].

Another Italian survey showed a decrease in the number of times sexual intercourse took place during quarantine, compared with before the period, in particular the surveyed people who had sexual intercourse more than twice a week (54.2 vs. 37.2%, $p < 0.01$) [30]. It was due to lack of privacy (43.2%) and lack of stimuli (40.9%) [30].

During a pandemic, many problems are faced by non-cohabitant partners.

Lopes showed that, in cohabiting couples, an improvement of the couple's sex life could be explained by the search for safety, intimacy, or by increasing the possibility of sexual intercourse [31]. In other cases, a worsening could be explained by opposite habits, and the search for compromises in respect of privacy and individuality [31]. The routine that was once taken for granted could become a source of stress. In these cases, sexual activity was certainly overshadowed. In unstable scenarios, men's libidos can be affected as much as women; desire may increase (as a relief valve, to seek immediate pleasure) or may be completely absent (loss of sense of security and stability). While partners who do not live together can adopt new sexual routines [31]. Furthermore, other studies have showed that external stressful events can provoke a decrease in sexual activity and satisfaction [30,31].

Deprivation of sexual activity could have very insidious psychological effects , particularly at a time when people are fragile, and their mental health is particularly strained [31]. Often, the participants reported being alone at home, away from their partners, and the rest of their families. Each participant faced the sad situation in a different way, particularly from a sexual standpoint. In literature, for example, the impact of social distancing on sexuality has been evaluated in men undergoing radical prostatectomy [32]. It is a surgical procedure that has a high impact on male sexuality. Depression, anxiety, and deprivation of sexual activity, during lockdown, can decrease the desire for sexual rehabilitation; the subject then enters a vicious circle in which his emotions are a cause and effect of his sexual problems.

Previous studies have evaluated and even recommended the use of alternative sexual practices, such as masturbation [13] or "virtual sex" via digital platforms, such as phone or video chat [31] or viewing pornography movies [13]. One Italian survey reported decreased masturbation activity due to poor privacy (46.4%) and lack of desire (34.7%) [30]

The limitations of our study include the lack of baseline data on sexual dysfunction, although this aspect was not the aim of this study. Furthermore, considering the lack of information about any comorbidities, it is not possible to adjust results obtained for potential factors affecting sexual habits.

Other limitations include the sample being very large; however, it may not be representative of the general population. We have not analyzed data on homosexuals and bisexuals due to too small a sample; we used only four items of MAT questionnaires, and we could not include subjects who did not have the internet. Future studies should consider this.

The strengths of our study include the sample size. To our knowledge, the present study is the first to evaluate the relationship between the COVID-19 pandemic and Italian female and male sexual functions by using standardized questionnaires, evaluating sexual dysfunctions, and analyzing people in this mean age (43.07 ± 12.5).

5. Conclusions

In conclusion, the lockdown and social distancing during the COVID-19 pandemic mostly improved couples' sex lives among cohabiting participants. The results of this research could be useful for interventions designed to help couples maintain sexual intimacy when they are not forced to spend more time together.

Supplementary Materials: The following are available online at https://www.mdpi.com/article/10.3390/jcm10071414/s1.

Author Contributions: E.I., E.C., F.T.: drafted the manuscript, data analysis, interpreted data, gave final approval of the version to be published; F.T., D.V., S.M., V.L.M., F.N., M.B. (Matteo Balzarro), V.M., R.B., F.M., M.B. (Marianna Bevacqua), A.L.P., E.A., M.G., M.T.F., G.D.R., E.F.A., P.S., C.B., L.G., A.G.: acquisition of data, gave final approval of the version to be published; G.R.: acquisition of data, English revision of the draft; E.I., F.T., G.R.: data analysis, gave final approval of the version to be published; E.I., F.T., G.R. interpretation of data, gave final approval of the version to be published;

E.C.: conception and design, general coordinator of the study, gave final approval of the version to be published. A.Z.: data curation, investigation, validfation, writing-review and editing, S.B.: data curation. All authors read and approved the final version of the manuscript.

Funding: This research received no external funding.

Institutional Review Board Statement: It was approved by the Local Bioethics Committee (University of Perugia) N. 44765.

Informed Consent Statement: Informed consent was obtained from all subjects involved in the study.

Data Availability Statement: Not applicable.

Conflicts of Interest: The authors declare no conflict of interest.

References

1. Ministero della Salute. Available online: www.salute.gov.it (accessed on 28 February 2021).
2. Agrò, E.F.; Farullo, G.; Balzarro, M.; Del Popolo, G.; Giannantoni, A.; Herms, A.; Marzi, V.L.; Musco, S.; Giammò, A.; Costantini, E. Triage of functional, female and neuro-urology patients during and immediately after the Covid-19 outbreak. *Minerva Urol. Nefrol.* **2020**, *72*, 513–515.
3. Chiancone, P.F. Fedelini Managing change in the urology department of a large hospital in Italy during the COVID-19 pan-demic. *Int. J. Urol.* **2020**, *27*, 820–822. [CrossRef] [PubMed]
4. Istituto Nazionale di Statistica. Available online: www.istat.it (accessed on 28 February 2021).
5. Forbes, M.K.; Krueger, R.F. The Great Recession and Mental Health in the United States. *Clin. Psychol. Sci.* **2019**, *7*, 900–913. [CrossRef]
6. Margerison-Zilko, C.; Goldman-Mellor, S.; Falconi, A.; Downing, J. Health impacts of the greatrecession: A critical review. *Curr. Epidemiol. Rep.* **2016**, *3*, 81–91. [CrossRef]
7. Carmassi, C.; Dell'Oste, V.; Pedrinelli, V.; Barberi, F.M.; Rossi, R.; Bertelloni, C.A.; Dell'Osso, L. Is Sexual Dysfunction in Young Adult Survivors to the L'Aquila Earthquake Related to Post-traumatic Stress Disorder? A Gender Perspective. *J. Sex. Med.* **2020**, *17*, 1770–1778. [CrossRef]
8. Bodenmann, G.; Atkins, D.C.; Schär, M.; Poffet, V. The association between daily stress and sexual activity. *J. Fam. Psychol.* **2010**, *24*, 271–279. [CrossRef]
9. Karney, B.R.; Bradbury, T.N. The longitudinal course of marital quality and stability: A review of theory, method, and research. *Psychol. Bull.* **1995**, *118*, 3–34. [CrossRef]
10. Huang, Y.; Zhao, N. Generalized anxiety disorder, depressive symptoms and sleep quality during COVID-19 outbreak in China: A web-based cross-sectional survey. *Psychiatry Res.* **2020**, *288*, 112954. [CrossRef] [PubMed]
11. Arafat, S.Y.; Alradie-Mohamed, A.; Kar, S.K.; Sharma, P.; Kabir, R. Does COVID-19 pandemic affect sexual behaviour? A cross-sectional, cross-national online survey. *Psychiatry Res.* **2020**, *289*, 113050. [CrossRef] [PubMed]
12. Jacob, L.; Smith, L.; Butler, L.; Barnett, Y.; Grabovac, I.; McDermott, D.; Armstrong, N.; Yakkundi, A.; Tully, M.A. Challenges in the Practice of Sexual Medicine in the Time of COVID-19 in the United Kingdom. *J. Sex. Med.* **2020**, *17*, 1229–1236. [CrossRef] [PubMed]
13. Cocci, A.; Giunti, D.; Tonioni, C. Love at the Time of the Covid-19 Pandemic: Preliminary Results of an Online Survey Con-ducted During the Quarantine in Italy. *Int. J. Impot. Res.* **2020**, *32*, 556–557. [CrossRef]
14. Filocamo, M.T.; Serati, M.; Marzi, V.L.; Costantini, E.; Milanesi, M.; Pietropaolo, A.; Polledro, P.; Gentile, B.; Maruccia, S.; Fornia, S.; et al. The Female Sexual Function Index (FSFI): Linguistic Validation of the Italian Version. *J. Sex. Med.* **2014**, *11*, 447–453. [CrossRef] [PubMed]
15. Cappelleri, J.C.; Rosen, R.C.; Smith, M.D.; Mishra, A.; Osterloh, I.H. Diagnostic evaluation of the erectile function domain of the international index of erectile function. *Urology* **1999**, *54*, 346–351. [CrossRef]
16. Bertoni, A.M.M.; Iafrate, R. Percezione del conflitto e soddisfazione coniugale: Un confronto tra mariti e mogli. *G. Ital. Psicol.* **2005**, *2005*, 171–189.
17. Fava, G.A.; Kellner, R.; Munari, F.; Pavan, L. The Hamilton Depression Rating Scale in normals and depressives. *Acta Psychiatr. Scand.* **1982**, *66*, 26–32. [CrossRef] [PubMed]
18. Rimé, B. Emotion Elicits the Social Sharing of Emotion: Theory and Empirical Review. *Emot. Rev.* **2009**, *1*, 60–85. [CrossRef]
19. Williams, W.C.; Morelli, S.A.; Ong, D.C.; Zaki, J. Interpersonal emotion regulation: Implications for affiliation, perceived support, relationships, and well-being. *J. Pers. Soc. Psychol.* **2018**, *115*, 224–254. [CrossRef]
20. Villani, M.T.; Morini, D.; Spaggiari, G.; Simoni, M.; Aguzzoli, L.; Santi, D. Spontaneous pregnancies among infertile couples during assisted reproduction lockdown for COVID-19 pandemic. *Andrology* **2021**. [CrossRef]
21. Lazarus, R.S.; Folkman, S. *Stress, Appraisal, and Coping*; Springer: Berlin/Heidelberg, Germany, 1984.
22. Maslow, A.H. *Motivation and Personality*; Harper & Row Publishers: New York, NY, USA, 1954.
23. Bahar, Y.; Faruk, O. Effect of the COVID-19 pandemic on female sexual behavior. *Int. J. Gynecol. Obstet.* **2020**, *150*, 98–102.

24. Hall, K.S.; Kusunoki, Y.; Gatny, H.; Barber, J. Stress Symptoms and Frequency of Sexual Intercourse Among Young Women. *J. Sex. Med.* **2014**, *11*, 1982–1990. [CrossRef]
25. Liu, S.; Han, J.; Xiao, D.; Ma, C.; Chen, B. A report on the reproductive health of women after the massive 2008 Wenchuan earthquake. *Int. J. Gynecol. Obstet.* **2010**, *108*, 161–164. [CrossRef]
26. Hamilton, L.D.; Meston, C.M. Chronic Stress and Sexual Function in Women. *J. Sex. Med.* **2013**, *10*, 2443–2454. [CrossRef]
27. Gilhooly, P.E.; Ottenweller, J.E.; Lange, G.; Tiersky, L.; Natelson, B.H. Chronic Fatigue and Sexual Dysfunction in Female Gulf War Veterans. *J. Sex Marital. Ther.* **2001**, *27*, 483–487. [CrossRef]
28. Hendrickx, L.; Gijs, L.; Enzlin, P. Prevalence rates of sexual difficulties and associated distress in heterosexual men and wom-en: Results from an internet survey in Flanders. *J. Sex Res.* **2013**, *51*, 112.
29. Nobre, P.J.; Pinto-Gouveia, J. Cognitions, Emotions, and Sexual Response: Analysis of the Relationship among Automatic Thoughts, Emotional Responses, and Sexual Arousal. *Arch. Sex. Behav.* **2007**, *37*, 652–661. [CrossRef] [PubMed]
30. Cito, G.; Micelli, E.; Cocci, A.; Polloni, G.; Russo, G.I.; Coccia, M.E.; Simoncini, T.; Carini, M.; Minervini, A.; Natali, A. The Impact of the COVID-19 Quarantine on Sexual Life in Italy. *Urology* **2021**, *147*, 37–42. [CrossRef] [PubMed]
31. Pereira Lopes, G.; Castro Vale, F.B.; Vieira, I.; da Silva Filho, A.; Abuhid, C.; Geber, S. COVID-19 and Sexuality: Reinventing Intimacy. *Arch. Sex Behav.* **2020**, *49*, 2735–2738. [CrossRef]
32. Chiancone, F.; Fabiano, M.; Fedelini, M.; Carrino, M.; Meccariello, C.; Fedelini, P. Preliminary evidences of the impact of social distancing on psychological status and functional outcomes of patients who underwent robot-assisted radical prostatectomy. *Central Eur. J. Urol.* **2020**, *73*, 265–268. [CrossRef]

Journal of
Clinical Medicine

Review

Analysis of the Clinical and Epidemiological Meaning of Screening Test for SARS-CoV-2: Considerations in the Chronic Kidney Disease Patients during the COVID-19 Pandemic

Francesca Martino [1,*], Gianpaolo Amici [2], Stefano Grandesso [3], Rosella Ferraro Mortellaro [2], Antonina Lo Cicero [2] and Giacomo Novara [4,*]

1 UO Nephrology, Dialysis, and Transplantation, San Bortolo Hospital, 36100 Vicenza, Italy
2 UO Nephrology and Dialysis, San Daniele del Friuli and Tolmezzo Hospital, ASUFC,
   33038 San Daniele del Friuli, Italy; amicig@tin.it (G.A.); rosella.ferraro-mortellaro@asufc.sanita.fvg.it (R.F.M.);
   antonina.locicero@asufc.sanita.fvg.it (A.L.C.)
3 Laboratory Medicine, Dolo-Mirano District, AULSS 3 Serenissima, 30100 Venezia, Italy;
   stefano.grandesso@aulss3.veneto.it
4 Department of Surgery, Oncology, and Gastroenterology, Urology Clinic University of Padua,
   35124 Padua, Italy
* Correspondence: francesca.martino.k@gmail.com (F.M.); giacomo.novara@unipd.it (G.N.);
   Tel.: +39-044-4753650 (F.M.); +39-049-8211250 (G.N.)

Abstract: The COronaVIrus Disease 19 (COVID-19) pandemic is an emerging reality in nephrology. In a continuously changing scenario, we need to assess our patients' additional risk in terms of attending hemodialysis treatments, follow-up peritoneal dialysis, and kidney transplant visits. The prevalence of severe acute respiratory syndrome coronavirus 2 (SARS-CoV-20 infection in the general population plays a pivotal role in estimating the additional COVID-19 risk in chronic kidney disease (CKD) patients. Unfortunately, local prevalence is often obscure, and when we have an estimation, we neglect the number of asymptomatic subjects in the same area and, consequently, the risk of infection in CKD patients. Furthermore, we still have the problem of managing COVID-19 diagnosis and the test's accuracy. Currently, the gold standard for SARS-CoV-2 detection is a real-time reverse transcription-polymerase chain reaction (rRT-PCR) on respiratory tract samples. rRT-PCR presents some vulnerability related to pre-analytic and analytic problems and could impact strongly on its diagnostic accuracy. Specifically, the operative proceedings to obtain the samples and the different types of diagnostic assay could affect the results of the test. In this scenario, knowing the local prevalence and the local screening test accuracy helps the clinician to perform preventive measures to limit the diffusion of COVID-19 in the CKD population.

Keywords: COVID-19; chronic kidney disease; screening text; accuracy

Citation: Martino, F.; Amici, G.; Grandesso, S.; Ferraro Mortellaro, R.; Lo Cicero, A.; Novara, G. Analysis of the Clinical and Epidemiological Meaning of Screening Test for SARS-CoV-2: Considerations in the Chronic Kidney Disease Patients during the COVID-19 Pandemic. *J. Clin. Med.* **2021**, *10*, 1139. https://doi.org/10.3390/jcm10051139

Academic Editor: Antonio Bellasi

Received: 22 January 2021
Accepted: 25 February 2021
Published: 9 March 2021

Publisher's Note: MDPI stays neutral with regard to jurisdictional claims in published maps and institutional affiliations.

1. Introduction

COronaVIrus Disease 19 (COVID-19) is a pandemic disease currently present in more of 200 countries globally [1], and it is caused by severe acute respiratory syndrome coronavirus 2 (SARS-CoV-2). COVID-19 can manifest in different ways from asymptomatic form to severe pneumonia and fatal multi-organ failure. The government in Italy and other countries decided to impose lockdown for an extended period to avoid a wide diffusion of COVID-19 and dilute the correlated need for severe cases of hospitalization. We started limiting unnecessary motion and promoting social distance, mask use, and hand cleaning in this context.

Chronic kidney disease (CKD) patients, especially those in stage 5D, have mandatory needs to attend hospital facilities visits or treatments. This condition per se could increase the risk of COVID-19 and should be carefully evaluated by nephrologists considering their local situation.

2. COVID-19 Risk in Chronic Kidney Patients

In the general population, the typical symptoms at onset are fever, dry cough, fatigue, and dyspnoea. Still, in some cases, the patients can present headaches, diarrhea, vomiting, abdominal pain, and dizziness [2,3]. Furthermore, physicians have reported other manifestations such as rash, eye abnormalities, and neurological and heart complications [2–4]. The clinical manifestations seem worse in elderly patients and those with comorbidities such as diabetes, hypertension, chronic kidney disease, chronic obstructive pulmonary disease (COPD), and chronic heart disease [5]. The presence of underlying kidney disease seems a risk factor for developing severe complications and appears to be associated with a higher mortality rate [6]. Specifically, a metanalysis including 1389 COVID-19 patients showed an odds ratio (OR) as high as 3 to have severe COVID-19 in the patients with previous CKD [7]. Additionally, CKD patients often suffer from hypertension, diabetes, and heart disease, which are consolidated risk factors for the deleterious progression of COVID-19 [5].

Generally, CKD patients present the same symptoms and signs of the general population [8]. On the basis of literature reports, we reported in Table 1 symptoms and laboratory features common in CKD patients with COVID-19 and their meaning in risk analysis for the development of severe complications such as acute respiratory distress syndrome (ARDS) and death.

Table 1. Significance of symptoms and signs of COronaVIrus Disease 19 (COVID-19) in chronic kidney disease (CKD) patients [9].

Symptoms/Signs	Increased Risk of ARDS	Increased Risk of Death
Cough	At onset ≈	At onset +
Fever	At onset +++	At onset +++
Shortness of breath	At onset +++	At onset ++
Gastrointestinal symptoms nausea vomiting diarrhea	Not significant	Not significant
Pharyngitis	Not significant	Not significant
Shortness of breath	Not significant	Not significant
Myalgia	At onset ++	Not significant
Blood examination	Not significant	Not significant
Lymphocytes decrease	Not significant	Not significant
Platelets decrease	Not significant	Not significant
C-RP increase	>50 mg/L +	>50 mg/L ++
AST/ALT increase	>50 U/L +	Not significant
LDH increase	Not significant	Not significant
Infiltrates at the chest X-ray	At onset +	Pneumonia ++

Footnotes: ARDS: acute respiratory distress syndrome, C-RP: C-reactive protein, AST: aspartate aminotransferase, ALT: alanine aminotransferase, LDH: lactate dehydrogenase, ≈: uncertain meaning, +: low risk, ++: average risk, +++: high risk.

Furthermore, the same considerations are substantially valid in kidney transplant patients, in whom the most common symptoms of COVID-19 onset were fever and dyspnea, followed by diarrhea and myalgia [10]. Specifically, in this class of patients, the mortality rate seems to be influenced by the age (OR 1.07), the respiratory rate at presentation >20 breaths/min (OR 6.88), and the kidney function evaluated by estimated Glomerular Filtrate Rate (OR 0.96).

The early phase presentation is not specific to COVID-19, making it difficult to recognize the exordia of disease and prevent diffusion and severe complications. Accordingly, nephrology and dialysis units adopted special programs to individualize potential COVID-19 patients.

At every dialysis session or nephrology consult before the facility access, healthcare workers provide a simple triage, evaluating the presence of symptoms and detecting the presence of high temperature and lower O_2 saturation [11,12]. In doubtful cases, patients are tested for SARS-CoV-2 and start quarantine until exclusion or confirmation of COVID-19 diagnosis. In COVID-19 diagnosis, CKD patients should be ideally transferred to a designated hospital or ward for COVID-19 patients if they need hospitalization. The in-hospital patients who require renal replacement therapy should be treated in an isolation room, and their healthcare workers should wear personal protective equipment (such as KF94 or N95 masks, gloves, goggles, or face shield, level D gown) when performing dialysis [13].

All previous procedures try to limit diffusion in the nephrology and dialysis unit. COVID-19 diffusion control shows its weakness in transmitting SARS-CoV-2 by asymptomatic people to fragile subjects in the general population [14] and potentially can affect CKD patients. However, the transmission by asymptomatic people seems to have a doubtful impact on dialysis patients as suggested by a Lombard study. This study showed a similar positive rate in real-time reverse transcription-polymerase chain reaction (rRT-PCR) in the hemodialysis unit where all patients were screened and in units where only symptomatic patients were screened [9]. This phenomenon could have more than an explanation if, on the one hand, hemodialysis patients could be higher susceptible to severe complications in most of the cases.

On the other hand, the rRT-PCR screening test could be less sensitive in asymptomatic patients with lower viral load. In any case, transmission by asymptomatic people seems to be a reasonable problem and limits our ability to prevent COVID-19 diffusion. Therefore, we can only take prophylactic measures such as: educating patients and healthcare workers about the personal protective dispositive (e.g., masks, and gloves) and social distance; preparing appropriate waiting rooms or resting areas; providing surgical masks and hand disinfection before entering the Hemodialysis (HD) unit [15–17].

In a recent survey promoted by the Società Italiana Nefrologia (SIN) [18] on 358 centers, the authors reported a prevalence of COVD-19 equal to 3.41%, 1.36%, and 0.87% in hemodialysis, peritoneal dialysis, and kidney transplant patients, respectively. Unfortunately, only 15% of centers performed at least one screening test on all patients. Furthermore, the authors reported a high death rate in CKD patients with SARS-CoV-2 infection: 49% of mortality in peritoneal dialysis patients, 37% of death in hemodialysis patients, and 25% in kidney transplant patients. On the basis of this preliminary report, in Italy, we see that the diffusion of COVID-19 in CKD patients seems higher than in the general population (as reported by the last updating of Istituto Superiore di Sanità (ISS), the rate of patients positive for SARS-CoV-2 was about 0.362% [218, 268/60, 317,000]). Furthermore, the crude mortality rate in CKD patients with COVID-19 is higher than in the general population, estimated in the same period by the last ISS report, at around 13.9% (30, 395/218, 268). The differences between CKD patients and the general population confirm our patients' fragility in terms of comorbidities and suggest a higher risk in the people who need frequent access to hospital facilities.

Furthermore, the Registry of the Spanish Society of Nephrology [8] confirms the same trend in COVID-19 dialysis and kidney transplant patients with a high rate of mortality (about 23%) and a high need for hospital admission (about 85%).

Finally, in a multicenter Turkish study on 1210 subjects, dialysis need, kidney transplant, and stage III-V CKD severely impacted on the patient prognosis, resulting in a higher rate of severe COVID-19 (25.4%, 21%, and 39.4%, respectively), and increased mortality (16.2%, 11.1%, and 28.4%, respectively) compared to the patients without kidney disease, for whom severe COVID-19 had a rate of about 8% and mortality of around 4% [19].

On the basis of previous considerations, we cannot consider the standard balance between the risk and the benefit enough for every procedure during the COVID-19 pandemic. Nephrologists have to know SARS-CoV-2 screening tests and their ability to predict COVID-19 to take adequate prophylactic measures to benefit each patient while considering the real risk. Specifically, SARS-CoV-2 screening test accuracy should be considered in patients who wait for a kidney transplant for the need to assess the balance between the risk and the benefit of the procedure in little time [20].

3. SARS-CoV-2 and Screening Test

3.1. SARS-CoV-2 Structure

SARS-CoV-2 is an RNA single-stranded virus belonging to the family of Coronaviridae, which is divided into four subfamilies: alfa, beta, gamma, and delta. SARS-CoV-2 belongs to beta-coronaviruses and shares at least 50% of its genome with other beta-coronaviruses SARS-CoV and Middle East respiratory syndrome coronavirus (MERS-CoV) members. During replication, SARS-CoV-2 produces 16 non-structural proteins and 6-9 structural and accessory proteins, such as spike (S), envelope (E), membrane (M), and nucleocapsid (N) [21]. Each protein is encoded by a corresponding gene, targeted as N, E, S, and RNA-dependent RNA polymerase (RdRp) genes. Figure 1 reports the SARS-CoV-2 structure and RNA sequences.

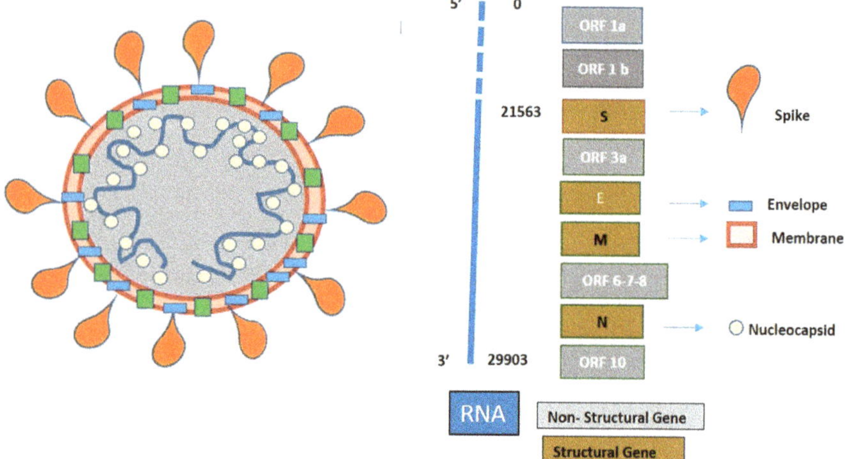

Figure 1. Schematic representation of severe acute respiratory syndrome coronavirus 2 (SARS-CoV-2) structure and genomic.

3.2. Screening Test Methods

Currently, the diagnosis of COVID-19 is confirmed by nucleic acid amplification tests (NAAT), such as real-time reverse transcription polymerase chain reaction (rRT-PCR) on the respiratory tract specimens [22].

After the extraction of RNA, rRT-PCR consists of a three-step procedure:

- Reverse transcription: a process where the enzyme reverse transcriptase converts RNA into complementary DNA (cDNA), which is suitable for PCR.
- Amplification of cDNA target sequences, which requires the presence of a polymerase enzyme and primer. The polymerase amplifies the cDNA sequence, while the primer identifies the specific sequences to amplify.

Detection, involves fluorescently labelling DNA oligonucleotides, which bind the primer and give a fluorescent signal at each amplification cycle. The fluorescence signal increases as more copies of DNA are produced; when the fluorescence arises to a certain threshold, the test is considered positive.

The World Health Organization (WHO) recommended E, N, and RdRp genes as molecular targets for first-line screening, as well as confirmatory tests on a nasopharyngeal or oropharyngeal swab, and on lower respiratory specimens (such as sputum, endotracheal aspirate, and bronchoalveolar lavage). Furthermore, at the website www.who.int/

emergencies/diseases/novel-coronavirus-2019/technical-guidance/laboratory-guidance published at 11 September 2020, accessed on 22 January 2021, the WHO provides technical guidance about the over 250 kits disposable on the market. Generally, the commercial kits detect the presence of two or three viral sequences. In the first case, identifying one gene is used as a screening test, while that of the second gene is used as a confirmatory test. In the latter case, a screening test is considered positive only when all genes are detected. Specifically, WHO suggested PCR amplification of the viral E gene as a screening test and amplification of the RdRp region of the orf1b gene as a confirmatory test. Afterwards, on 12 March 2020, the European Centre for Disease Prevention and Control (ECDC) specified no absolute need for a confirmatory test. Specifically, in lower transmission countries, a confirmatory test is always required. In contrast, in the countries with high transmission, a confirmatory test's performance is only required when the first result is technically not interpretable, or the RT-PCR cycle threshold value is above 35 [23].

3.3. Screening Test Accuracy

Despite the gold standard's endorsement, rRT-PCR is not flawless, and it has shown accuracy problems, which can lead to underrating SARS-CoV-2 infection.

The COVID-19 pandemic is supposed to be a high-prevalence disease with serious consequences for the patients. In this scenario, a screening test should have high sensitivity with a lower false-negative rate. Precisely, a higher rate of false negatives limits the ability of screening tests to recognize the patients with COVID-19 and consequently increases the likelihood to delay the medical care of COVID-19 patients. This aspect is dangerous for CKD patients, who showed high susceptibility to develop serious consequences after SARS-CoV-2 infection. Additionally, a high rate of false negatives is critical during a pandemic because it does not allow one to follow the recommendations to limit the diffusion in the community without extra cost for the health system. Specifically, in CKD patients, a high false-negative rate increases the risk of dissemination in the hospital facility.

Unfortunately, rRT-PCR's sensitivity rate was estimated to be around 66–80% [24] in a Chinese study of 1014 patients. On the basis of this report, we see that the accuracy of the rRT-PCR test in the diagnosis of COVID-19 seems to be weak and related to different types of issues. In an exciting review by Lippi et al. [25], the authors reported the two kinds of laboratory problems: preanalytical (such as inadequate procedures for collection, handling, transport and storage, collection of inappropriate or unsuitable material, presence of interfering substances) and analytical (such as testing outside the diagnostic window, active viral recombination, use of inadequately validated assays, insufficient harmonization, and instrument malfunctioning). All these procedural matters result in a high risk of a false-negative test.

Furthermore, the commercially available diagnostics kits in rRT-PCR have different characteristics, mainly due to the viral region investigated and the limit of detection (LoD). Noticeably, the higher the LoD, the more risk of false negatives. In Table 2, we present the characteristics of some of the kits mainly used in Italy. Finally, when the clinical picture is strongly suspected for COVID-19 infection, and the swab is repeatedly negative [26,27], and it may be appropriate to carry out a serological investigation to search for IgM and IgG [26].

Finally, we want to highlight how not only the rate of false negatives but also the rate of false positive results negatively influences the management of vulnerable patients. In the first case, as we emphasized in the previous paragraph, there is a high likelihood of contagious between the patients with potentially devastating consequences for the relatively small CKD communities (patients, health workers, and support personnel). Conversely, in the case of false positive tests, there is a waste of resources for the surveillance and the management of standard care, as well as concomitant psycho-physical stress in patients that is highly proven by their basal health conditions.

Table 2. Commercially available diagnostics kits mainly used in Italy in real-time reverse transcription polymerase chain reaction (rRT-PCR) with gene target and limit of detection.

Company (Assay Name)	Gene Target	LoD	Specimen Types	Approval
Abbott Diagnostics (*ID NOW COVID-19*)	RdRp	125 copies/mL	Nasal, throat, NPS	FDA (US)
Abbott Molecular (Abbott RealTime SARS-CoV-2 EUA Test)	RdRp, N	100 virus copies/mL	NPS, OPS, nasal swab, BAL	FDA (US) CE-IVD
Cepheid (*Xpert Xpress SARS CoV-2*)	N2, E	250 copies/mL	NPS, OPS, nasal, mid-turbinate swab, nasal wash/aspirate	FDA (US), Health Canada, Australia, Singapore, Philippines, Brazil
DiaSorin Molecular (LIAISON MDX)	ORF1ab, S gene	NPS: 500 copies/mL, Nasal swab: 242 copies/mL	Nasal swab, NPS, nasal wash/aspirate, BAL	CE-IVD
Tib Molbiol (*Modular DX kit SARS-CoV-2*)	E	1–10 copies/reaction	OPS, NPS	RUO (research use only)
Roche Molecular System (Cobas 6800 SARS-CoV-2)	ORF-1a/b, E	1000 RNA genome equivalents/mL	NPS, OPS	US-FDA, CE-IVD
Seegene (Allplex 2019-nCoV Assay)	RdRp, N, E	100 RNA copies/rxn	NPS, NPA, OPS, sputum, BAL	Korea (Korea CDC), US-FDA, CE-IVD
bioMerieux (ARGENE SARS-CoV-2 R-GENE)	RdRp, N, E	380 genomic copies/mL	NPS	RUO (research use only)

Legend: LoD: limit of detection, RnRp: RNA-dependent RNA polymerase, N: nucleocapsid, E: envelope, ORF: open reading frame, EUA: Emergency Use Authorization, BAL: bronchoalveolar lavage, NPA: nasopharyngeal aspirate, NPS: nasopharyngeal swab, OPS: oropharyngeal swab, FDA: Food and Drug Administration, CE-IVD: European Conformity In-Vitro Diagnostic, CDC: Centers for Disease Prevention and Control.

3.4. Specimen Type

Between the preanalytical issues, the most debated argument is the type of specimen. One of the first reports about COVID-19 described a significant difference in the screening test's sensibility related to the kind of specimen [28]. Specifically, bronchoalveolar lavage fluid seemed to have the best accuracy with a rate of positive equal to 93%, sputum with a rate of 72%, nasal swabs with a rate of 63%, and finally pharyngeal swabs with a rate of 32%.

It seems accepted that the specimen derived from the upper respiratory tract shows its weakness compared with the low respiratory tract, especially during the symptomatic phase. Nasopharyngeal swab seems more suitable than oropharyngeal swab, which appears to have a higher rate of false negatives, as reported by Wang et al. in a comparative study on about 350 patients [29] and by Mohammadi in a recent meta-analysis [30]. Furthermore, saliva (a clear, slightly alkaline liquid secreted into the mouth by the salivary glands and mucous glands) seems to have the same reliability as nasopharyngeal swabs [31,32] and better reliability of oropharyngeal swabs [33], and thus it should be considered as an alternative specimen in the diagnosis of COVID-19 in symptomatic patients. Finally, sputum sampling (fluid coughed up and expectorated from the mouth, composed of saliva and discharges from the respiratory passages such as mucus and phlegm) seems to have higher sensitivity to nasopharyngeal swab. Likely, if other studies support its better sensitivity, in the future, we should consider the sputum as a preferred specimen in diagnosing and monitoring COVID-19 [30].

Furthermore, over the types of specimen, we have to consider the timing of collection. In the week before symptom onset, the viral load could be very low and likely inadequate for the detection by rRT-PCR. Consequently, in this phase of COVID-19, the screening test could have a high likelihood to have false-negative results [34]. As reported, the higher viral loads are detected soon after symptom onset [30] and can persist in throat swabs for more than 30 days [35,36]. The specimens' types show different accuracy profiling in various phases of COVID-19, likely related to viral load, suggesting the preferable kind of sample and operative conditions, as reported in Table 3.

Table 3. COVID-19 disease phase and sample site recommendation.

Sample Sites	Asymptomatic Phase	Onset of the Symptomatic Phase	Symptomatic Phase	Convalescence Phase
Naso-pharyngeal swabs	Unclear	Highly recommended Detection rate: 80%	Recommended Detection rate: 59%	Recommended Detection rate: 36%
Oro-pharyngeal swabs	Unclear	Highly recommended Detection rate: 75%	Not recommended Detection rate: 35%	Not recommended Detection rate: 12%
Saliva collection	Unclear	Highly recommended Detection rate: 82.2%	Unclear	Unclear
Sputum collection	Unclear	Highly recommended Detection rate: 98%	Highly recommended Detection rate: 69%	Not recommended Detection rate: 46%
Bronco-alveolar lavage	Unclear/not recommend	Unclear/not recommended	Highly recommended in intubated patients Detection rate: 94%	Not recommend
Fecal/anal swabs	Not recommend	Not recommended Detection rate: 48%	Not recommended	Recommended Detection rate: 73%

Consequently, we have to prefer upper respiratory tract specimens in the incubation period, such as nasopharyngeal swabs saliva/sputum collection. While in the symptomatic period, we have to choose the lower respiratory tract specimens (such as bronchoalveolar lavage fluid) in critical patients who require intubation. Finally, during convalescence, we suggest adding fecal/anal swab to the standard nasopharyngeal swab [37].

3.5. Statistical Insight on Screening Test

In general, any test has different performances in different settings or applications. In the COVID-19 screening test case, different disease prevalence can lead to surprisingly different interpretations of tests, even with the same value of sensitivity and specificity. Table A1 reports in synthesis the common statistical knowledge and calculations about test performance evaluation. Positive and negative predictivity value has a key role in interpreting a single test result in a clinical setting because it suggests to a physician whether the test results are trustable. In other words, positive predictive value (PPV) offers the probability of having an ill patient when the result of the test is positive, and negative predictive value (NPV) tells of the probability of having healthy patients when the result is negative.

Specifically, in the COVID-19 pandemic, we observed a different prevalence of the disease in the same population, likely related to the seasonal period and the use of adequate prophylactic measures, which have an obvious impact on the interpretation of the screening test for vulnerable patients, such as CKD patients. Unfortunately, it is not always simple to individualize the real prevalence in different areas considering the variable rate of asymptomatic people, the number of screened people, and the frequency of the screening. Despite these considerations, we suggest optimizing the available information such as the number of COVID-19 patients in the local hospitals, as well as the local reports by the authorities to understand the trend in COVID-19 diffusion.

Tables A2–A5 and Figure 2 report some examples relative to the accuracy of the screening tests for COVID-19, which only have an explicative role. The reported examples are extrapolated from the sensibility and specificity described in some studies [24,27] and show how COVID-19 prevalence and the value of sensitivity and specificity can impact the test's interpretation. Unfortunately, nasopharyngeal swabs' real sensitivity and specificity are only partially known with a value of sensibility of 66–80% and specificity of 90–95%.

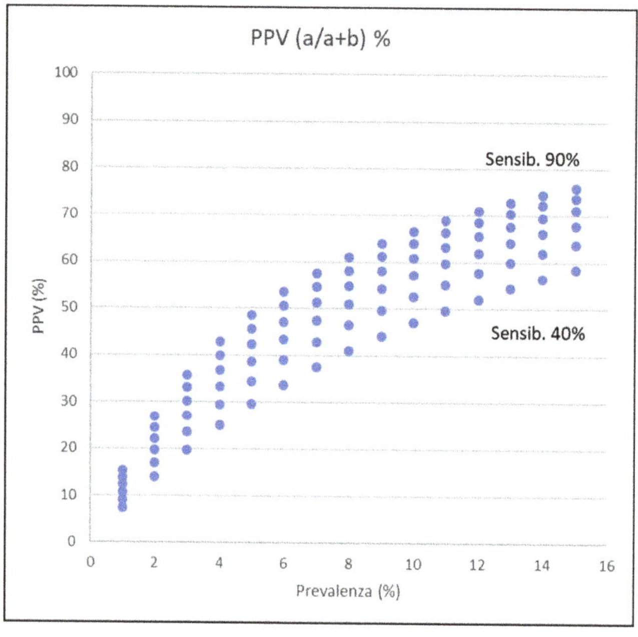

(a)

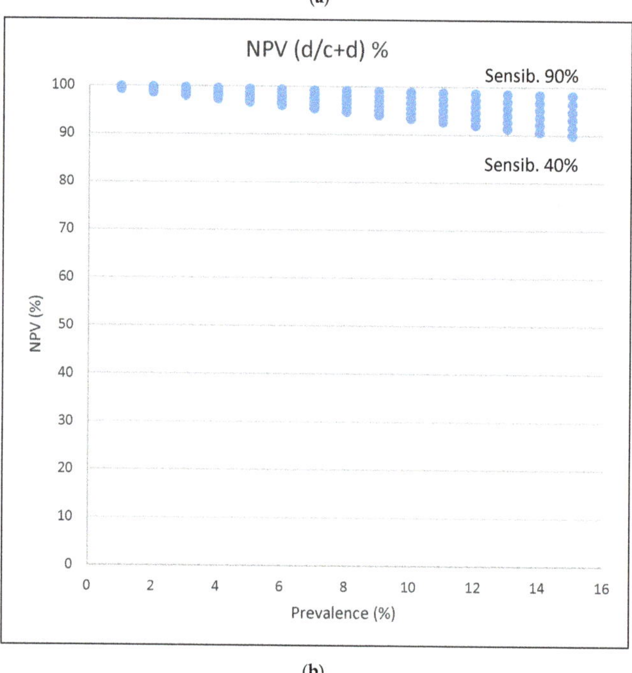

(b)

Figure 2. Graphic representation of positive predictive value (PPV) and negative predictive value (NPV) variations with different percentage disease prevalence (*X*-axis, from 1 to 15%), sensibility variation from 40 to 90%, and fixed specificity at 95%. a: true positive, b: false positive, c: false negative, d: true negative. (**a**) PPV values show an increase with increasing disease prevalence. (**b**) NPV values are instead decreasing with disease prevalence increase.

In high prevalence conditions, and optimal sensitivity and specificity of nasopharyngeal swab, a single result gives high values of predictivity, both positive and negative. However, in a low prevalence situation, the same test with the same sensitivity and specificity gives significantly lower positive predictive values. If we consider oropharyngeal swabs, that show low sensitivity and good specificity, resulting in a poor positive predictive value (largely not useful for screening purposes) and good negative predicted value. Specifically, we have worse PPV and better NPV in a low prevalence situation.

4. Conclusions

COVID-19 has been shown to be very risky in CKD patients in terms of the development of serious consequences, such as acute respiratory distress syndrome and death. During the COVID-19 pandemic, the screening test for SARS-CoV-2 is considered the gold standard for the diagnosis. Unfortunately, different issues such as the sensibility/specificity of the single test, the period of performing, the type of specimen, and the prevalence of disease could strongly impact on the interpretation of the test and its reliability. In order to reduce the contagious between the patients, nephrologists have to carefully manage the results of the screening test for SARS-CoV-2, considering the suboptimal sensitivity of the test and the relevant likelihood of false-negative results. In this scenario, promoting extensive use of protective measures (such as the personal protective dispositive, social distance, and a limitation of simultaneous access to nephrology facilities) seems a reasonable approach. When possible, considering the local resources, intensifying the number of samples for each patient could be theoretically recommended to overcome the accuracy issue of the screening test. Finally, we suggest considering the anal swab to readmit CKD patients who have SARS-CoV-2 infection to the hospital facilities.

Author Contributions: Conceptualization, F.M., G.A. and S.G.; methodology, F.M. and G.N.; software, formal analysis, all authors.; investigation, all authors.; resources, all authors.; data curation, all authors.; writing—original draft preparation, F.M. and G.A.; writing—review and editing, G.N.; visualization, all authors.; supervision, G.N.; project administration, R.F.M. and A.L.C. All authors have read and agreed to the published version of the manuscript.

Funding: The study did not receive any kind of funding. Specifically, this research received no external funding.

Institutional Review Board Statement: Institutional Review Board Statement: Ethical review and approval were waived for this study due to the nature of the review.

Informed Consent Statement: Not applicable.

Data Availability Statement: Not applicable.

Conflicts of Interest: The authors declare no conflict of interest.

Appendix A.

Appendix A.1. Insight Box 1

Table A1. Definitions and calculation details about the evaluation of tests performances.

Test		Disease		
		Present	Absent	
	positive	a	b	a+b
	negative	c	d	c+d
		a+c	b+d	N

Definitions:

Sensitivity is the ability to designate a subject with the disease as positive correctly -> a/a+c.

Specificity is the ability to designate a healthy subject as negative correctly -> d/b+d.

Positive predictive value (PPV) is the probability that a subject with a positive test has the disease -> a/a+b. PPV is mostly useful in a clinical setting because it predicts the disease's likelihood when the test is positive.

Negative predictive value (NPV) is the probability that the subject has no disease given a negative test result -> d/c+d. PPV is mostly useful in a clinical setting because it predicts the likelihood to be healthy in a patient with the test is negative.

False-negative is the proportion of ill subjects who are misclassified as healthy by test -> c.

False-positive is the proportion of healthy subject who are misclassified as ill by test -> b.

Accuracy is the proportion of true results, either true positive or true negative, in a population. It measures the degree of veracity of a diagnostic test on a condition. -> a+d/N.

Prevalence is the proportion of subjects who have a disease in a population and represents the a priori probability of selecting a person with a disease in the population randomly understudy -> a+c/N. PPV and NPV depend on the disease's prevalence, while Sensitivity and Specificity are intrinsic characteristics of the test.

A screening test is a medical test to assess the likelihood of having a particular disease, and its principal aim is to detect potential disease indicators. Consequently, it should have the high sensibility and lower false negative. While the purpose of a diagnostic test is to establish the presence (or absence) of disease as a basis for treatment decisions in symptomatic or screen-positive individuals, it should have high specificity and a lower rate of false positive.

Appendix A.2. Insight Box 2: Tests Performance Examples in Different Scenarios of COVID-19 Pandemic

Table A2. Prevalence of COVID-19 equal to 9.5% with sensibility equal 84.2% and specificity equal 98.9%.

		Covid-19		
		Present	**Absent**	
Saliva	positive	16	2	18
	negative	30	179	182
		19	181	200

The numbers reported are based on the sensibility, and specificity reported by Pasomsub et al. [32]. PPV = 16/18 = 88.9% -> means when the test is positive; the likelihood to have a COVID-19 patient is 88.9%. NPV = 179/182 = 98.4% -> means when the test is negative; the likelihood to have a NO COVID-19 patient is 98.4%.

Table A3. Prevalence of COVID-19 equal to 0.36% with sensibility equal 84.2% and specificity equal 98.9%.

		Covid-19		
		Present	**Absent**	
Saliva	positive	60	22	82
	negative	12	1906	1918
		72	1928	2000

The numbers reported are based on the sensibility and specificity reported by Pasomsub et al [32]. PPV=60/82=73.2% -> means when the test is positive; the likelihood to have a COVID-19 patient is 73.2%. NPV= 1906/1918= 99.4% -> means when the test is negative, the likelihood to have a NO COVID-19 patient is 99.4%.

Table A4. Prevalence of COVID-19 equal to 9.5% with sensibility equal 27% and specificity equal 84.9%.

		Covid-19		
		Present	Absent	
Oro-pharyngeal swabs	positive	5	27	32
	negative	14	154	168
		19	181	200

The numbers reported are based on the sensibility and specificity reported by Wang et al [29]. PPV = 5/32 = 15.6% -> means when the test is positive; the likelihood to have a COVID-19 patient is 15.6%. NPV = 154/168 = 91.7% -> means when the test is negative; the likelihood to have a NO COVID-19 patient is 91.7%.

Table A5. Prevalence of COVID-19 equal to 0.36% with sensibility equal 27% and specificity equal 84.9%.

		Covid-19		
		Present	Absent	
Oro-pharyngeal swabs	positive	19	292	311
	negative	53	1636	1689
		72	1928	2000

The numbers reported are based on the sensibility and specificity reported by Wang et al [29]. PPV = 19/311 = 6.1% -> means when my test is positive the likelihood to have a COVID-19 patient is 6.1%. NPV = 1636/1689 = 96.9% -> means when my test is negative the likelihood to have a NO COVID-19 patient is 96.9%.

The previous tables are extracted from a study published, which reported sensitivity and specificity of saliva and oropharyngeal swab considering as reference nasopharyngeal swab, but unfortunately the real sensitivity and specificity of nasopharyngeal swabs remain uncertain.

References

1. Sokouti, M.; Sadeghi, R.; Pashazadeh, S.; Eslami, S.; Sokouti, M.; Ghojazadeh, M.; Sokouti, B. Comparative Global Epidemiological Investigation of SARS-CoV-2 and SARS-CoV Diseases Using Meta-MUMS Tool Through Incidence, Mortality, and Recovery Rates. *Arch. Med Res.* **2020**, *51*, 458–463. [CrossRef] [PubMed]
2. Wang, D.; Hu, B.; Hu, C.; Zhu, F.; Liu, X.; Zhang, J.; Wang, B.; Xiang, H.; Cheng, Z.; Xiong, Y.; et al. Clinical Characteristics of 138 Hospitalized Patients With 2019 Novel Coronavirus-Infected Pneumonia in Wuhan, China. *JAMA* **2020**, *323*, 1061–1069. [CrossRef]
3. Huang, C.; Wang, Y.; Li, X.; Ren, L.; Zhao, J.; Hu, Y.; Zhang, L.; Fan, G.; Xu, J.; Gu, X.; et al. Clinical features of patients infected with 2019 novel coronavirus in Wuhan, China. *Lancet* **2020**, *395*, 497–506. [CrossRef]
4. Tammaro, A.; Adebanjo, G.A.R.; Parisella, F.R.; Pezzuto, A.; Rello, J. Cutaneous manifestations in COVID-19: The experiences of Barcelona and Rome. *J. Eur. Acad. Dermatol. Venereol.* **2020**, *34*, e306–e307. [CrossRef]
5. He, R.; Lu, Z.; Zhang, L.; Fan, T.; Xiong, R.; Shen, X.; Feng, H.; Meng, H.; Lin, W.; Jiang, W.; et al. The clinical course and its correlated immune status in COVID-19 pneumonia. *J. Clin. Virol.* **2020**, *127*, 104361. [CrossRef] [PubMed]
6. Cheng, Y.; Luo, R.; Wang, K.; Zhang, M.; Wang, Z.; Dong, L.; Li, J.; Yao, Y.; Ge, S.; Xu, G. Kidney disease is associated with in-hospital death of patients with COVID-19. *Kidney Int.* **2020**, *97*, 829–838. [CrossRef]
7. Henry, B.M.; Lippi, G. Chronic kidney disease is associated with severe coronavirus disease 2019 (COVID-19) infection. *Int. Urol. Nephrol.* **2020**, *52*, 1193–1194. [CrossRef]
8. Sánchez-Álvarez, J.E.; Pérez Fontán, M.; Jiménez Martín, C.; Blasco Pelícano, M.; Cabezas Reina, C.J.; Sevillano Prieto, Á.M.; Melilli, E.; Crespo Barrios, M.; Macía Heras, M.; Del Pino, Y.; et al. SARS-CoV-2 infection in patients on renal replacement therapy. Report of the COVID-19 Registry of the Spanish Society of Nephrology (SEN). *Nefrologia* **2020**, *40*, 272–278. [CrossRef] [PubMed]
9. Alberici, F.; Delbarba, E.; Manenti, C.; Econimo, L.; Valerio, F.; Pola, A.; Maffei, C.; Possenti, S.; Lucca, B.; Cortinovis, R.; et al. A report from the Brescia Renal COVID Task Force on the clinical characteristics and short-term outcome of hemodialysis patients with SARS-CoV-2 infection. *Kidney Int.* **2020**, *98*, 20–26. [CrossRef]
10. Craved, P.; Mothi, S.S.; Azzi, Y.; Haverly, M.; Farouk, S.S.; Pérez-Sáez, M.J.; Redondo-Pachón, M.D.; Murphy, B.; Florman, S.; Cyrino, L.G.; et al. COVID-19 and kidney transplantation: Results from the TANGO International Transplant Consortium. *Am. J. Transplant.* **2020**, *20*, 3140–3148. [CrossRef]
11. Scarpioni, R.; Manini, A.; Valsania, T.; De Amicis, S.; Albertazzi, V.; Melfa, L.; Ricardi, M.; Rocca, C. Covid-19 and its impact on nephropathic patients: The experience at Ospedale "Guglielmo da Saliceto" in Piacenza. *G. Ital. Nefrol.* **2020**, *37*, 1–5.
12. Meijers, B.; Messa, P.; Ronco, C. Safeguarding the Maintenance Hemodialysis Patient Population during the Coronavirus Disease 19 Pandemic. *Blood Purif.* **2020**, *49*, 259–264. [CrossRef]

13. Park, H.C.; Kim, D.H.; Yoo, K.D.; Kim, Y.G.; Lee, S.H.; Yoon, H.E.; Kim, D.K.; Kim, S.N.; Kim, M.S.; Jung, Y.C.; et al. Korean Society of Nephrology COVID-19 Task Force Team. Korean clinical practice guidelines for preventing transmission of coronavirus disease 2019 (COVID-19) in hemodialysis facilities. *Kidney Res. Clin. Pract.* **2020**, *39*, 145–150. [CrossRef]

14. Gandhi, M.; Yokoe, D.S.; Havlir, D.V. Asymptomatic Transmission, the Achilles' Heel of Current Strategies to Control Covid-19. *N. Engl. J. Med.* **2020**, *382*, 2158–2160. [CrossRef] [PubMed]

15. Ikizler, T.A. COVID-19 and Dialysis Units: What Do We Know Now and What Should We Do? *Am. J. Kidney Dis.* **2020**, *76*, 1–3. [CrossRef] [PubMed]

16. Lee, J.J.; Lin, C.Y.; Chiu, Y.W.; Hwang, S.J. Take proactive measures for the pandemic COVID-19 infection in the dialysis facilities. *J. Formos. Med. Assoc.* **2020**, *119*, 895–897. [CrossRef] [PubMed]

17. Yang, X.H.; Sun, R.H.; Zhao, M.Y.; Chen, E.Z.; Liu, J.; Wang, H.L.; Yang, R.L.; Chen, D.C. Expert recommendations on blood purification treatment protocol for patients with severe COVID-19: Recommendation and consensus. *Chronic Dis. Transl. Med.* **2020**, *6*, 106–114. [CrossRef]

18. Quintaliani, G.; Reboldi, G.; Di Napoli, A.; Nordio, M.; Limido, A.; Aucella, F.; Messa, P.; Brunori, G. Italian Society of Nephrology COVID-19 Research Group. Exposure to novel coronavirus in patients on renal replacement therapy during the exponential phase of COVID-19 pandemic: Survey of the Italian Society of Nephrology. *J. Nephrol.* **2020**, *33*, 725–736. [CrossRef]

19. Ozturk, S.; Turgutalp, K.; Arici, M.; Odabas, A.R.; Altiparmak, M.R.; Aydin, Z.; Cebeci, E.; Basturk, T.; Soypacaci, Z.; Sahin, G.; et al. Mortality analysis of COVID-19 infection in chronic kidney disease, haemodialysis and renal transplant patients compared with patients without kidney disease: A nationwide analysis from Turkey. *Nephrol. Dial. Transplant.* **2020**, *35*, 2083–2095. [CrossRef]

20. Martino, F.; Plebani, M.; Ronco, C. Kidney transplant programmes during the COVID-19 pandemic. *Lancet Respir. Med.* **2020**, *8*, e39. [CrossRef]

21. Rehman, M.F.U.; Fariha, C.; Anwar, A.; Shahzad, N.; Ahmad, M.; Mukhtar, S.; Farhan, U.L.; Haque, M. Novel coronavirus disease (COVID-19) pandemic: A recent mini review. *Comput. Struct. Biotechnol. J.* **2021**, *19*, 612–623. [CrossRef]

22. Ahn, D.G.; Shin, H.J.; Kim, M.H.; Lee, S.; Kim, H.S.; Myoung, J.; Kim, B.T.; Kim, S.J. Current Status of Epidemiology, Diagnosis, Therapeutics, and Vaccines for Novel Coronavirus Disease 2019 (COVID-19). *J. Microbiol. Biotechnol.* **2020**, *30*, 313–324. [CrossRef]

23. Novel Coronavirus Disease 2019 (COVID-19) Pandemic: Increased Transmission in the EU/EEA and the UK – Sixth Update. Available online: https://www.ecdc.europa.eu/sites/default/files/documents/RRA-sixth-update-Outbreak-of-novel-coronavirus-disease-2019-COVID-19.pdf (accessed on 12 March 2020).

24. Ai, T.; Yang, Z.; Hou, H.; Zhan, C.; Chen, C.; Lv, W.; Tao, Q.; Sun, Z.; Xia, L. Correlation of Chest CT and RT-PCR Testing for Coronavirus Disease 2019 (COVID-19) in China: A Report of 1014 Cases. *Radiology* **2020**, *296*, E32–E40. [CrossRef]

25. Lippi, G.; Simundic, A.M.; Plebani, M. Potential preanalytical and analytical vulnerabilities in the laboratory diagnosis of coronavirus disease 2019 (COVID-19). *Clin. Chem. Lab. Med.* **2020**, *58*, 1070–1076. [CrossRef]

26. Cao, G.; Tang, S.; Yang, D.; Shi, W.; Wang, X.; Wang, H.; Li, C.; Wei, J.; Ma, L. The potential transmission of SARS-CoV-2 from patients with negative RT-PCR swab tests to others: Two related clusters of COVID-19 outbreak. *Jpn. J. Infect. Dis.* **2020**. [CrossRef] [PubMed]

27. Winichakoon, P.; Chaiwarith, R.; Liwsrisakun, C.; Salee, P.; Goonna, A.; Limsukon, A.; Kaewpoowat, Q. Negative Nasopharyngeal and Oropharyngeal Swabs Do Not Rule Out COVID-19. *J. Clin. Microbiol.* **2020**, *58*, e00297-e20. [CrossRef]

28. Bwire, G.M.; Majigo, M.V.; Njiro, B.J.; Mawazo, A. Detection profile of SARS-CoV-2 using RT-PCR in different types of clinical specimens: A systematic review and meta-analysis. *J. Med. Virol.* **2020**. [CrossRef] [PubMed]

29. Wang, X.; Tan, L.; Wang, X.; Liu, W.; Lu, Y.; Cheng, L.; Sun, Z. Comparison of nasopharyngeal and oropharyngeal swabs for SARS-CoV-2 detection in 353 patients received tests with both specimens simultaneously. *Int. J. Infect. Dis.* **2020**, *94*, 107–109. [CrossRef] [PubMed]

30. Mohammadi, A.; Esmaeilzadeh, E.; Li, Y.; Bosch, R.J.; Li, J. SARS-CoV-2 Detection in Different Respiratory Sites: A Systematic Review and Meta-Analysis. *medRxiv* **2020**. [CrossRef]

31. McCormick-Baw, C.; Morgan, K.; Gaffney, D.; Cazares, Y.; Jaworski, K.; Byrd, A.; Molberg, K.; Cavuoti, D. Saliva as an Alternate Specimen Source for Detection of SARS-CoV-2 in Symptomatic Patients Using Cepheid Xpert Xpress SARS-CoV-2. *J. Clin. Microbiol.* **2020**, *58*, e01109-20. [CrossRef]

32. Pasomsub, E.; Watcharananan, S.P.; Boonyawat, K.; Janchompoo, P.; Wongtabtim, G.; Suksuwan, W.; Sungkanuparph, S.; Phuphuakrat, A. Saliva sample as a non-invasive specimen for the diagnosis of coronavirus disease 2019: A cross-sectional study. *Clin. Microbiol. Infect.* **2020**. [CrossRef]

33. Lukassen, S.; Chua, R.L.; Trefzer, T.; Kahn, N.C.; Schneider, M.A.; Muley, T.; Winter, H.; Meister, M.; Veith, C.; Boots, A.W.; et al. SARS-CoV-2 receptor ACE2 and TMPRSS2 are primarily expressed in bronchial transient secretory cells. *EMBO J.* **2020**, *39*, e105114. [CrossRef]

34. Zou, L.; Ruan, F.; Huang, M.; Liang, L.; Huang, H.; Hong, Z.; Yu, J.; Kang, M.; Song, Y.; Xia, J.; et al. SARS-CoV-2 Viral Load in Upper Respiratory Specimens of Infected Patients. *N. Engl. J. Med.* **2020**, *382*, 1177–1179. [CrossRef]

35. Yang, J.R.; Deng, D.T.; Wu, N.; Yang, B.; Li, H.J.; Pan, X.B. Persistent viral RNA positivity during the recovery period of a patient with SARS-CoV-2 infection. *J. Med. Virol.* **2020**. [CrossRef] [PubMed]

36. Chen, Y.; Li, L. SARS-CoV-2: Virus dynamics and host response. *Lancet Infect. Dis.* **2020**, *20*, 515–516. [CrossRef]

37. Song, F.; Zhang, X.; Zha, Y.; Liu, W. COVID-19: Recommended sampling sites at different stages of the disease. *J. Med. Virol.* **2020**. [CrossRef] [PubMed]

Journal of
Clinical Medicine

Article

Triage of Patients Suspected of COVID-19 in Chronic Hemodialysis: Eosinophil Count Differentiates Low and High Suspicion of COVID-19

Romain Vial [1,†], Marion Gully [1,†], Mickael Bobot [1,2], Violaine Scarfoglière [1], Philippe Brunet [1,2], Dammar Bouchouareb [1], Ariane Duval [3], He-oh Zino [1], Julien Faraut [1], Océane Jehel [1], Yaël Berdad-Haddad [4], Stéphane Burtey [1,2], Pierre-André Jarrot [2,5], Guillaume Lano [1,2] and Thomas Robert [1,6,*]

[1] Centre of Nephrology and Renal Transplantation, Hôpital de la Conception, CHU de Marseille, 13005 Marseille, France; Romain.VIAL@ap-hm.fr (R.V.); marion.gully@ap-hm.fr (M.G.); mickael.bobot@ap-hm.fr (M.B.); violaine.scarfogliere@ap-hm.fr (V.S.); philippe.brunet@ap-hm.fr (P.B.); dammar.bouchouareb@ap-hm.fr (D.B.); Heoh.zino@ap-hm.fr (H.-o.Z.); julien.faraut@ap-hm.fr (J.F.); oceane.jehel@ap-hm.fr (O.J.); stephane.burtey@ap-hm.fr (S.B.); guillaume.lano@ap-hm.fr (G.L.)
[2] C2VN, Aix-Marseille University, INSERM 1263, INRAe, 13005 Marseille, France; Pierre.JARROT@ap-hm.fr
[3] Association des Dialysés Provence et Corse, 13009 Marseille, France; Ariane.DUVAL@ap-hm.fr
[4] Hematology Laboratory, Hôpital de la Conception, CHU de Marseille, 13005 Marseille, France; Yael.BERDA@ap-hm.fr
[5] Department of Internal Medicine and Clinical Immunology, CHU de Marseille, Hôpital de la Conception, 13005 Marseille, France
[6] MMG, Bioinformatics & Genetics, Aix-Marseille Université, UMR_S910, 13004 Marseille, France
* Correspondence: thomas.robert@ap-hm.fr
† Romain Vial and Marion Gully contributed equally to the work.

Citation: Vial, R.; Gully, M.; Bobot, M.; Scarfoglière, V.; Brunet, P.; Bouchouareb, D.; Duval, A.; Zino, H.; Faraut, J.; Jehel, O.; et al. Triage of Patients Suspected of COVID-19 in Chronic Hemodialysis: Eosinophil Count Differentiates Low and High Suspicion of COVID-19. *J. Clin. Med.* **2021**, *10*, 4. https://doi.org/10.3390/jcm10010004

Received: 16 November 2020
Accepted: 17 December 2020
Published: 22 December 2020

Publisher's Note: MDPI stays neutral with regard to jurisdictional claims in published maps and institutional affiliations.

Abstract: Background: Daily management to shield chronic dialysis patients from SARS-CoV-2 contamination makes patient care cumbersome. There are no screening methods to date and a molecular biology platform is essential to perform RT-PCR for SARS-CoV-2; however, accessibility remains poor. Our goal was to assess whether the tools routinely used to monitor our hemodialysis patients could represent reliable and quickly accessible diagnostic indicators to improve the management of our hemodialysis patients in this pandemic environment. Methods: In this prospective observational diagnostic study, we recruited patients from La Conception hospital. Patients were eligible for inclusion if suspected of SARS-CoV-2 infection when arriving at our center for a dialysis session between March 12th and April 24th 2020. They were included if both RT-PCR result for SARS-CoV-2 and cell blood count on the day that infection was suspected were available. We calculated the area under the curve (AUC) of the receiver operating characteristic curve. Results: 37 patients were included in the final analysis, of which 16 (43.2%) were COVID-19 positive. For the day of suspected COVID-19, total leukocytes were significantly lower in the COVID-19 positive group (4.1 vs. 7.4 G/L, $p = 0.0072$) and were characterized by lower neutrophils (2.7 vs. 5.1 G/L, $p = 0.021$) and eosinophils (0.01 vs. 0.15 G/L, $p = 0.0003$). Eosinophil count below 0.045 G/L identified SARS-CoV-2 infection with AUC of 0.9 [95% CI 0.81—1] ($p < 0.0001$), sensitivity of 82%, specificity of 86%, a positive predictive value of 82%, a negative predictive value of 86% and a likelihood ratio of 6.04. Conclusions: Eosinophil count enables rapid routine screening of symptomatic chronic hemodialysis patients suspected of being COVID-19 within a range of low or high probability.

Keywords: hemodialysis; COVID-19; eosinophil

1. Introduction

Severe Acute Respiratory Syndrome related-Coronavirus 2 (SARS-CoV-2) infection, also called Coronavirus Disease-19 (COVID-19) is a viral infection caused by a ribonucleic acid (RNA) virus of the coronavirus family. Since it was first described in Wuhan, China, this disease has become a global pandemic, and by April 29th 2020 more than three million people had been infected and more than 200,000 had died.

Chronic dialysis patients are a vulnerable group at high risk of SARS-CoV-2 contamination with at least one comorbidity–such as hypertension, being elderly and diabetes–associated with COVID-19 mortality [1,2]. First, dialysis patients are overexposed to the risk of disease transmission for logistical reasons (regular presence at health care facilities, repeated trips by ambulance or taxi and physical proximity of patients during hemodialysis) and have difficulties with respect to social distancing. Second, it is essential to be able to quickly diagnose affected dialysis patients in order to prevent the spread of the disease within the ward and to protect the dialysis population in each center. The workflow of chronic dialysis patients can be quickly stretched in the context of COVID-19. The implementation of a clinical triage of patients upon their arrival in the dialysis center makes it possible to identify patients suspected of infection. Dialysis centers have therefore set up COVID-19 isolation zones to limit the risk of transmission to non-suspect patients while waiting for real-time reverse-transcriptase polymerase chain reaction (RT-PCR) results. Each suspension creates stress and puts a strain on the organization of the dialysis center waiting the final diagnosis [3,4].

Diagnosis is based on nasopharyngeal real-time RT-PCR, for which the feasibility and timeliness depend on the capacities of each center. In any case, however, this procedure does not permit classification of the patient as COVID-19 positive or negative in less than 4 h in addition with the risk of false negative [5]. Chest computed tomography (CT) scans can screen for low or high suspicion of COVID-19 but these are not available at all dialysis centers [6]. Some biological parameters, such as ferritin, lymphocyte and eosinophil count, have been studied for screening the patient for a low or high suspicion of COVID-19 but no studies have been conducted in dialysis patients to date [7]. Patients undergoing dialysis receive a weekly or monthly schedule of biological monitoring [8]. The results of a blood count and standard biochemical analyses are available in less than 2 h.

We hypothesize that anomalies in the biological report on the day of suspected COVID infection, compared to the monthly report for a patient on their arrival in the dialysis center, can be identified in order to quickly determine a low or high suspicion of COVID-19 in less than 2 h.

2. Methods

2.1. Setting

In the context of the global pandemic of SARS-CoV-2, we have set up clinical screening for SARS-CoV-2 infection when patients arrive at the dialysis center of the Hôpital de la Conception, Assistance Publique–Hôpitaux de Marseille (APHM), Marseille, France. We prospectively collected data from patients identified as suspects during this screening between March 12th and April 24th 2020. The suspected cases were all tested for the SARS-CoV-2 virus by nasopharyngeal real-time RT-PCR to determine whether they were COVID positive or negative. Positive RT-PCR were confirmed twice times. Presence of one of the following symptoms at arrival in the dialysis unit suggested SARS-CoV-2 infection: fever, cough, dyspnea, rhinorrhea, headache, asthenia, anosmia, ageusia, diarrhea, nausea and/or vomiting, myalgia, confusion. The data included in this study was anonymized, approved according to General Data Protection Regulation and registered at the Health Data Portal and Data Protection Commission of APHM under the references PADS-20-154 and 2020-58. The patients were provided with oral information about this study.

2.2. Participants

The inclusion criteria in the study were: nasopharyngeal real-time RT-PCR assay for SARS-CoV-2 infection and complete blood count (CBC) on the same day. The exclusions criteria were: age < 18 years, patients under corticosteroid treatment, chemotherapy within the last three months, recent acute stress (severe trauma, major surgery, epileptic seizure, myocardial infarction in the previous month) and active hematological disease.

Patients who did not have a CBC on the previous routine monthly workup or whose initial nasopharyngeal real-time RT-PCR had not been analyzed at the APHM laboratory were excluded. The COVID-19 patients have been reported in another accepted publication [9].

2.3. Data Source/Measurement
2.3.1. Epidemiological and Clinical Data

From electronic medical records we collected the following data: demographic, clinical, laboratory results, nucleic acid test results. Baseline patient characteristics were collected from electronic medical records: age, gender, body mass index (BMI), comorbidities (initial nephropathy, vascular access, history of immunosuppression or kidney transplantation, heart failure, coronaropathy, peripheral artery disease, arrhythmia, chronic respiratory disease, diabetes, cancers, hypertension and smoking) and their significant treatments such as angiotensin-converting enzyme inhibitors (ACEI), angiotensin receptor blockers (ARB), vitamin K antagonist, calcium channel blockers, beta blockers, aspirin, clopidogrel, statins, non-steroidal anti-inflammatory drugs (NSAIDs), iron supplementation and erythropoietin in dialysis.

2.3.2. Laboratory Procedures

Methods for laboratory confirmation of SARS-CoV-2 infection: one virology laboratory was responsible for SARS-CoV-2 detection in respiratory specimens using real-time RT-PCR methods. Throat-swab specimens were obtained for SARS-CoV-2 RT-PCR in the dialysis unit. The system targeted the envelope protein (E)-encoding gene, as described previously [10]. RT-PCR was considered negative over a 34-cycle threshold (CT) value.

Routine blood examinations were CBC by an automated cell counter and serum biochemical tests (electrolyte, albuminemia, C-reactive protein [CRP]). We collected the routine monthly blood test monitoring (CBC, electrolyte, albuminemia, CRP) (results from March) for hemodialysis patients.

2.3.3. Statistical Analysis

Continuous and categorical variables were presented as median (interquartile range [IQR]) and n (%), respectively. Sensitivity and specificity, as well as positive and negative predictive values, were calculated.

We used the Mann–Whitney U test, χ^2 test, or Fisher's exact test to compare differences between negative and positive COVID-19 where appropriate. All tests were two-tailed.

Unconditional logistic regression analysis was used to determine whether each variable was an independent factor in COVID-19 diagnosis. Covariates for the multivariate logistic regression analysis were selected based on a p-value < 0.05 in a univariate analysis. Variables were considered significant if $p < 0.05$, and the results are presented as odds ratio with 95% confidence intervals (CIs). Diagnostic accuracy for COVID-19 was assessed using the receiver operating characteristic area under curve (ROC AUC). Cut-off values showing the greatest accuracy were determined using sensitivity/specificity. All statistical analyses performed with the Prism 8 (GraphPad Software Inc., La Jolla, CA, USA).

3. Results
3.1. Patient Characteristics

60 patients were tested for SARS-CoV-2 RNA detection by nasopharyngeal swab at the Hôpital de la Conception, APHM. After excluding 23 patients according to the non-inclusion criteria, 37 were included in the final analysis (flowchart, Figure 1). Among the 37 patients included, 39 real-time RT-PCR tests were performed (2 patients were screened twice) with a peak at week 3 (Figure 2). 21 patients were negative for the SARS-CoV-2 RT-PCR and 16 were positive (Table 1). 22 RT-PCR were negative and 17 were positive. The median age was 72 years (IQR 54.5–79), with a median BMI of 23.3 kg/m^2, and most of the patients were male (Table 1). Hypertension was the most represented comorbidity (86.5%), followed by atrial fibrillation (32.4%) and diabetes (30.6%) and only 4 patients had

chronic respiratory disease (10.8%) (Table 1). Antihypertensive treatments, particularly ACEI and ARB, are detailed in Table 1. At baseline, corresponding to the March blood sample, patients had normal white blood cell count, and none of these clinical or biological data differed between the COVID-19 positive or negative patient groups (Table 1).

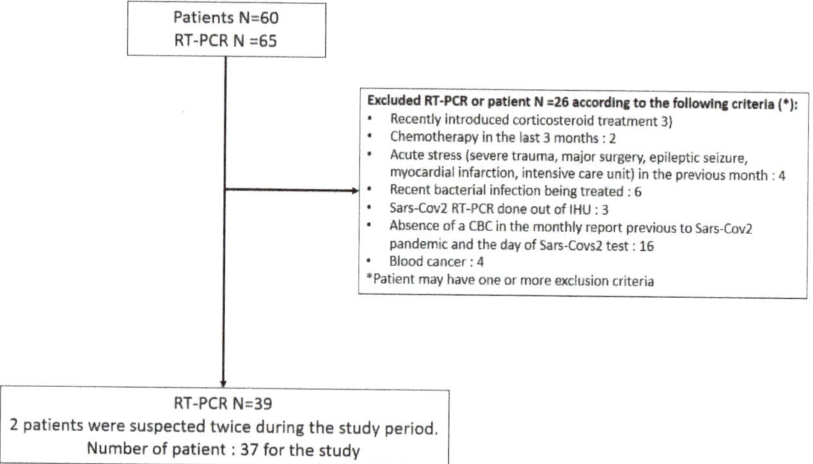

Figure 1. Flowchart. IHU, Institut Hospitalo-universitaire-Méditerranée Infection; RT-PCR, reverse-transcriptase polymerase chain reaction.

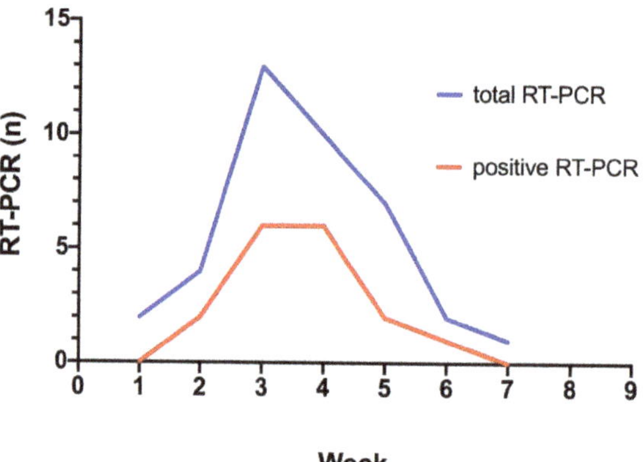

Figure 2. Evolution of the number of reverse-transcriptase polymerase chain reaction (RT-PCRs) testing for SARS-CoV-2 performed between March 12 and April 24 2020. The blue line represents the evolution of the number of RT-PCR performed between week 1 and 8. The red line represents the evolution of the number of positive RT-PCR sent during week 1 and 8.

The day of suspected COVID-19, total leukocytes were significantly lower in the COVID-19 positive group (4.1 vs. 7.4 G/L $p = 0.0072$). The white blood cell count was characterized by lower neutrophils (2.7 vs. 5.1 G/L, $p = 0.021$) and eosinophils (0.01 vs. 0.15 G/L, $p = 0.0003$). The remaining biological variables were not significantly different (Table 2). Compared to their baseline biological status, neutrophils from the COVID-19 negative group increased significantly the day of the COVID-19 suspicion (3.6 vs. 5.1 G/L, $p = 0.008$). For the COVID-19 positive group, lymphocytes (1.2 vs. 0.8 G/L, $p = 0.001$) and eosinophils decreased significantly (0.18 vs. 0.01 G/L, $p < 0.0001$). In both groups, we observed an increase in CRP (6.2 vs. 13, $p = 0.02$; 2.3 vs. 34.2, $p = 0.001$) (Table 3).

Table 1. Patients baseline characteristics.

Variable [a]	Total *n* = 37	Negative COVID *n* = 21	Positive COVID *n* = 16	*p* Value
Male	24 (64.9)	12 (57.2)	12 (75)	0.31
Age (years)	72 (54.5–79)	72 (48–83.5)	71 (55–80.9)	0.79
Weight (kg)	63.5 (57.5–70.6)	64.5 (58–69.3)	63.5 (55.6–81.8)	0.32
BMI (kg/m^2)	23.3 (20.6–24.6)	23.7 (20.8–24.5)	23.3 (19.8–25.7)	0.92
Nephropathy:				-
Glomerular	11 (29.7)	5 (23.8)	6 (37.5)	
Vascular	8 (21.6)	5 (23.8)	3 (18.6)	
Tubular	8 (21.6)	7 (33.3)	1 (6.3)	
Genetic	2 (5.5)	1 (4.8)	1 (6.3)	
Not determined	8 (21.6)	3 (14.3)	5 (31.3)	
Vascular access:				
Fistula	26 (70.3)	15 (71.4)	11 (68.8)	1
Central catheter	11 (29.7)	6 (28.6)	5 (31.2)	
Immunosuppression	11 (29.6)	7 (33.3)	4 (28.6)	1
History of graft kidney	7 (18.9)	4 (19.1)	3 (20)	1
Comorbidities				
Hypertension	32 (86.5)	19 (90.5)	13 (81.3)	0.63
Congestive heart failure	6 (12.5)	4 (19.5)	2 (12.5)	0.68
Coronary heart disease	7 (18.9)	2 (9.5)	5 (31.2)	0.2
Peripheral vascular disease	7 (18.9)	3 (14.3)	4 (25)	0.44
Cardiac arrhythmia	12 (32.4)	8 (40)	4 (25)	0.48
Chronic respiratory disease	4 (10.8)	3 (14.3)	1 (6.3)	0.62
Diabetes	11 (30.6)	7 (35)	4 (25)	0.72
Cancer	4 (10.8)	1 (4.8)	3 (18.8)	0.3
Smoker	5 (13.9)	4 (19.5)	1 (6.3)	0.35
Medication:				
ACE inhibitors	6 (16.2)	2 (9.5)	4 (25)	0.17
ARBs	5 (13.5)	3 (14.3)	2 (12.5)	1
Beta blocker	12 (32.4)	6 (28.6)	6 (37.5)	0.73
Calcium channel blockers	10 (27.0)	6 (28.6)	4 (25)	1
Diuretic	1 (2.7)	0	1 (6.3)	0.43
Aspirin	12 (32.4)	7 (33.3)	5 (31.3)	1
Clopidogrel	3 (8.1)	2 (9.5)	1 (6.3)	1
VK	9 (24.3)	6 (28.6)	3 (18.8)	0.7
Statin drug	5 (13.5)	1 (4.8)	4 (25)	0.14
Steroids	0	0	0	-
ASEs	27 (73)	17 (61.9)	10 (38.1)	0.38
March biological values				
Leukocyte (G/L)	5.9 (4.6–6.5)	5.8 (4.6–7.9)	5.9 (4.7–6.2)	0.54
Neutrophil (G/L)	3.6 (2.8–4.6)	3.6 (3.1–5.3)	3.6 (2.4–4.2)	0.44
Lymphocyte (G/L)	1 (0.75–1.45)	1 (0.6–1.2)	1.2 (0.9–1.6)	0.16
Monocyte (G/L)	0.6 (0.4–0.8)	0.6 (0.45–0.9)	0.5 (0.4–0.7)	0.29
Eosinophil (G/L)	0.17 (0.01–0.33)	0.17 (0.1–0.3)	0.18 (0.11–0.35)	0.44
Platelet (G/L)	189 (143–249)	198 (149–258)	179 (138–235)	0.71
Hemoglobin (g/dL)	11.2 (10.4–11.6)	11.2 (10–11.6)	11.2 (10.6–11.8)	0.78
CRP (mg/L)	4.7 (1.35–8.45)	6.2 (2.8–9.9)	2.3 (0.7–38.8)	0.95
Albumin (g/L)	39.3 (37.2–42)	39 (37.2–42)	39.8 (37.2–42.3)	0.47
Potassium (mmol/L)	4.5 (4.1–5.3)	4.3 (4–5)	5.1 (4.1–5.7)	0.2

[a] For quantitative variables, values are expressed as median (interquartile range). For qualitative variables, values are expressed as *n* (%). ACE, angiotensin-converting enzyme; ARBs, angiotensin-receptor blockers; ASEs, Erythropoiesis-stimulating agents; BMI, body mass index; CRP, C-reactive protein; VK, vitamin K antagonist; -, No statistic test were performed.

Table 2. Variables the day of COVID-19 suspicion.

Variables [a]	Negative COVID	Positive COVID	*p* Value
Leukocyte (G/L)	7.4 (4.9–10.4)	4.1 (3.3–7.1)	0.0072
Neutrophil (G/L)	5.1 (3.3–8.1)	2.7 (2.2–5.6)	0.021
Lymphocyte (G/L)	0.85 (0.57–1.22)	0.8 (0.55–1.05)	0.29
Monocyte (G/L)	0.80 (0.47–0.92)	0.5 (0.3–0.8)	0.099
Eosinophil (G/L)	0.15 (0.06–0.43)	0.01 (0–0.04)	0.0003
Platelet (G/L)	201 (152–255)	162 (118–185)	0.077
Hemoglobin (g/dL)	11.1 (10.5–11.6)	11.2 (10.4–12.4)	0.81
CRP (mg/L)	13 (3.3–65.5)	34.2 (15.9–72.8)	0.57
Albumin (g/L)	40.5 (36.1–42.4)	37.6 (34.2–41.5)	0.24
Potassium (mmol/L)	4.6 (4.15–5.55)	4.8 (4.1–5.1)	0.32

[a] For quantitative variables, values are expressed as median (interquartile range). CRP, C-reactive protein.

Table 3. Biological comparison between March monitoring and the suspicion day.

Variable [a]	Negative RT-PCR			Positive RT-PCR		
	Monthly Assessment	Suspicion Day	*p* Value	Monthly Assessment	Suspicion Day	*p* Value
Leukocyte (G/L)	5.8 (4.6–7.9)	7.4 (4.9–10.4)	0.09	5.9 (4.7–6.2)	4.1 (3.3–7.1)	0.16
Neutrophil (G/L)	3.6 (3.1–5.3)	5.1 (3.3–8.1)	0.008	3.6 (2.4–4.2)	2.7 (2.2–5.6)	0.84
Lymphocyte (G/L)	1 (0.6–1.2)	0.85 (0.57–1.22)	0.30	1.2 (0.9–1.6)	0.8 (0.55–1.05)	0.001
Monocyte (G/L)	0.6 (0.45–0.9)	0.8 (0.47–0.92)	0.48	0.5 (0.4–0.7)	0.5 (0.3–0.8)	0.82
Eosinophil (G/L)	0.17 (0.1–0.3)	0.15 (0.06–0.43)	0.23	0.18 (0.11–0.35)	0.01 (0–0.04)	<0.0001
Platelet (G/L)	198 (149–258)	201 (152–255)	0.64	179 (138–235)	162 (118–185)	0.004
Hemoglobin (g/dL)	11.2 (10–11.6)	11.1 (10.5–11.6)	0.24	11.2 (10.6–11.8)	11.2 (10.4–12.4)	0.71
CRP (mg/L)	6.2 (2.8–9.9)	13 (3.3–65.5)	0.02	2.3 (0.7–38.8)	34.2 (15.9–72.8)	0.001
Albumin (g/L)	39 (37.2–4)	40.5 (36.1–42.4)	0.38	39.8 (37.2–42.3)	37.6 (34.2–41.5)	0.07
Potassium (mmol/L)	4.3 (4–5)	4.6 (4.15–5.6)	0.16	5.1 (4.1–5.7)	4.8 (4.1–5.1)	0.27

[a] For quantitative variables, values are expressed as median (interquartile range). CRP, C-reactive protein; RT-PCR, reverse-transcriptase polymerase chain reaction.

3.2. Diagnostic Accuracy of Eosinopenia

Eosinopenia was observed in 14 out of 17 the SARS-CoV-2 RT-PCR positive group versus 3 out of 22 the SARS-CoV-2 RT-PCR negative group. ROC AUC for the detection of SARS-CoV-2 was 0.9 (0.81–1) ($p < 0.0001$). The highest diagnostic accuracy was observed for eosinophil count cut-off at 0.045 G/L. The eosinopenia diagnostic performance for SARS-CoV-2 infection showed a sensitivity of 82%, specificity of 86%, a positive predictive value of 82%, a negative predictive value of 0.86% and a likelihood ratio of 6.04 (Table 4 and Figures 3 and 4).

Table 4. Diagnostic performance for eosinopenia and RT-PCR the day of suspicion.

Effect Size	Value	95% CI
Sensitivity	0.82	0.59 to 0.94
Specificity	0.86	0.67 to 0.95
Positive Predictive Value	0.82	0.59 to 0.94
Negative Predictive Value	0.86	0.67 to 0.95
Likelihood Ratio	6.04	

CI, confidence interval; RT-PCR, reverse-transcriptase polymerase chain reaction.

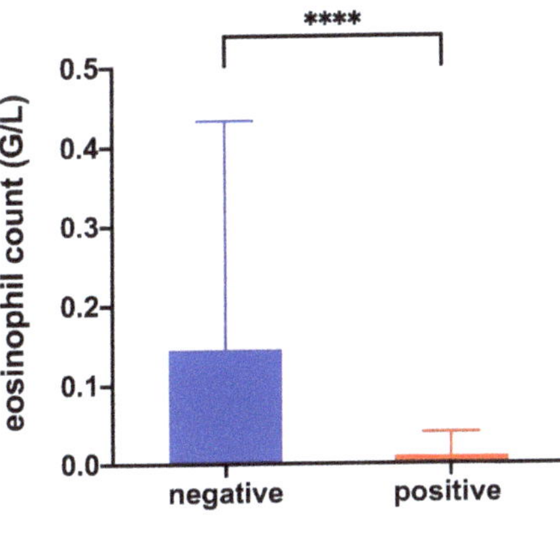

Figure 3. Comparison of the eosinophil level when performing a cell blood count on the day of a COVID-19 suspicion. Values represented are median and IQR. Mann–Whitney test. RT-PCR, reverse-transcriptase polymerase chain reaction; ****, represent p value < 0.0001.

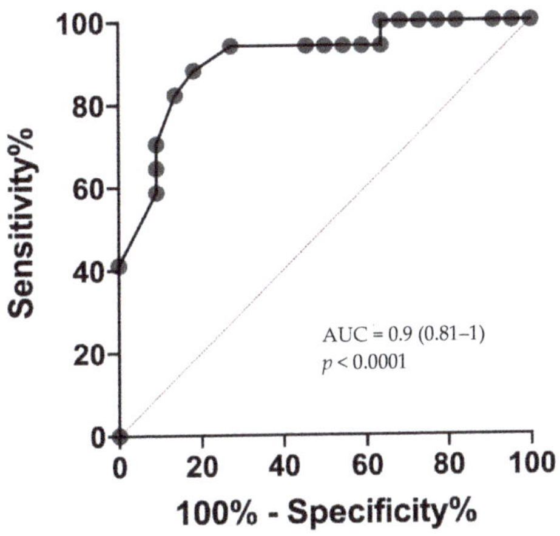

Figure 4. Receiver operating characteristic (ROC) curve of eosinophil count showing specificity and sensitivity for COVID-19 diagnosis. AUC, area under the ROC curve.

4. Discussion

This is the first study to show that development of eosinopenia can differentiate low and high COVID-19 suspicion in chronic dialysis patients, with high diagnostic accuracy.

Eosinophils are found predominantly in tissues, with a smaller fraction found in circulation. The half-life of the eosinophil in the peripheral blood of normal individuals is approximately 18 h, with an average blood transit time of around 26 h, similar to that of neutrophils. Eosinophil is a cell that is principally present in extravascular sites in quantities several hundred times greater than those in peripheral blood. Circulating cells reflect only those that transit between the blood marrow and their final extravascular functional destination. During certain acute inflammatory or immune responses, a time lag between the migration of circulating eosinophils to the tissue where the immune response takes place and the induction of eosinophil synthesis and marrow emigration is observed [11]. This leads to the development of either eosinopenia or delayed blood eosinophilia, or both, and may explain the presence of eosinopenia in patients with COVID-19 disease. Our results indicate that early development of eosinopenia could reflect a powerful acute inflammatory and immune response triggered by SARS-CoV-2 infection. The role of eosinopenia in COVID-19 remains unclear and may be multifactorial. Whether the acquired eosinopenia associated with COVID-19 directly contributes to the disease course or is a marker of severe disease has not yet been determined. However, evaluation of the eosinophilic blood count represents a useful tool to manage early SARS-CoV-2 suspicion for the dialysis patients and in deciding to promptly isolate a patient from the other dialyzed patient in the center.

In this study, eosinophil count reliably discriminated between patients with and those without COVID-19 with an AUC of 0.9 by using a cutoff of 0.045 G/L within 24 h of the suspected diagnosis. The discrimination between low and high COVID-19 suspicion is a challenge and clinically relevant. We have not tested the role of eosinopenia in comparison with influenzae in our cohort because we did not observe co-infection in our center. This point has been studied by Andreozzi et al. in a letter to editor and raise the point that complete eosinopenia is a common finding in both COVID-19 and Seasonal Influenza infections. Eosinopenia is a potential biological indicator of either Influenza or SARS-COV-2 infections. However, complete eosinopenia should raise the suspicion of a COVID-19 infection outside of the flu season [12]. We believe that detection of eosinopenia is of interest to detect more quickly COVID-19 infection and promptly isolated the patient in the dialysis center. The ability to identify high COVID-19 suspicion with an inexpensive, widely available, point-of-care test has important practical implications, particularly in the efforts to screen hemodialysis patients during their thrice weekly management. Interestingly, classical markers of inflammation such as CRP are not discriminating in our population, as COVID negative patients were subject to an infectious process during screening. In COVID-19 dialysis patients, we found a similar tendency to lymphopenia and thrombocytopenia as in the general COVID-19 population [13]. In our study the onset of lymphopenia is non-discriminatory, probably because this population is characterized by acquired immune deficiencies secondary to the uremic stage. In contrast, we found no tendency to hypokalemia in our COVID-19 dialysis patients, which can be partly explained by end-stage renal disease.

Molecular biology and chest CT scans, if available, with subsequent results, take more than 12 h for most chronic hemodialysis centers. In contrast, CBC is a routine procedure in these centers. Results are obtained within one hour, allowing for the identification of low or high suspicion soon after the arrival of the dialysis patient. In our study, more than 50% of suspected patients included in the final analysis tested negative for COVID. The diagnostic approach was based on the result of the SARS-CoV-2 real-time RT-PCR, where reporting time did not permit ruling out or confirming the diagnosis of COVID-19 by the end of the dialysis session. The presence of eosinopenia could thus make it possible to classify patients as low or high suspicion and help the clinician to improve the diagnostic process.

Our study has a small sample size, which is its main limitation. We have deliberately excluded patients with factors that might have interfered with eosinophil interpretation and thus represent another limitation of our study. In addition, we benefited from pre-pandemic biological characteristics, allowing us to show the development of eosinopenia using the COVID-19 diagnostic. The diagnostic accuracy of our study needed to be externally validated in another data set of hemodialysis patient and represent a limitation of the study.

These encouraging results lead us to believe that it is possible to carry out a systematic screening of patients based on the CBC at each weekly check-up. It would be interesting to assess if eosinopenia enable to identify asymptomatic patients and reduce contagiousness in our vulnerable population.

5. Conclusions

To conclude, the detection of eosinopenia enables rapid triage of symptomatic chronic hemodialysis patients into low or high COVID-19 suspicion groups when they arrive at the dialysis session. The ability to identify high COVID suspicion with an inexpensive, widely available, point-of-care test, has important practical implications, particularly for early hemodialysis isolation to avoid spread of SARS-CoV-2 throughout a center. This low cost triage tool is of particular interest in the coming months, especially for low-income countries with limited access to RT-PCR and chest CT scans [14,15].

Author Contributions: R.V., M.G., V.S., GL., and T.R. designed the study; R.V., M.G. and T.R. analyzed the data; R.V., O.J. and J.F. made the tables; R.V., T.R., S.B., H-o.Z., P.B., Y.B.-H., P.-A.J., M.B., D.B., A.D., and G.L., drafted and revised the paper; All authors have read and agreed to the published version of the manuscript.

Funding: No funding was obtained for this study.

Institutional Review Board Statement: The study was conducted according to the guidelines of the Declaration of Helsinki, and approved by the Institutional Review Board (or Ethics Committee) of the Health Data Portal of Assistance Publique—Hôpitaux de Marseille (protocol code 2020-58 and 2020 May).

Informed Consent Statement: Informed consent was obtained from all subjects involved in the study.

Data Availability Statement: The data presented in this study are available on request from the corresponding author T.R.

Acknowledgments: We would like to thank Gilles Kaplanski who noticed that eosinophils were greatly decreased in COVID patients hospitalized in the unit for which he was the head at the Hôpital de la Conception.

Conflicts of Interest: The authors declare that they have no competing interests.

Abbreviations

ACEI	Angiotensin-converting enzyme inhibitors
APHM	Assistance Publique–Hôpitaux de Marseille
ARB	Angiotensin receptor blockers
BMI	Body mass index
CBC	Complete blood count
COVID-19	Coronavirus Disease-19
CRP	C-reactive protein
CT	Chest computed tomography
IQR	Interquartile range
NSAIDs	Non-steroidal anti-inflammatory drugs
SARS-CoV-2	Severe acute respiratory syndrome related-coronavirus 2
ROC AUC	Receiver operating characteristic area under curve
RNA	Ribonucleic acid
RT-PCR	Real-time reverse-transcriptase polymerase chain reaction

References

1. Zhou, F.; Yu, T.; Du, R.; Fan, G.; Liu, Y.; Liu, Z.; Xiang, J.; Wang, Y.; Song, B.; Gu, X.; et al. Clinical course and risk factors for mortality of adult inpatients with COVID-19 in Wuhan, China: A retrospective cohort study. *Lancet* **2020**, *395*, 1054–1062. [CrossRef]
2. Cheng, Y.; Luo, R.; Wang, K.; Zhang, M.; Wang, Z.; Dong, L.; Li, J.; Yao, Y.; Ge, S.; Xu, G. Kidney disease is associated with in-hospital death of patients with COVID-19. *Kidney Int.* **2020**, *97*, 829–838. [CrossRef] [PubMed]

3. Burgner, A.; Ikizler, T.A.; Dwyer, J.P. COVID-19 and the Inpatient Dialysis Unit: Managing Resources During Contingency Planning Pre-Crisis. *Clin. J. Am. Soc. Nephrol.* **2020**, *15*, 720–722. [CrossRef] [PubMed]

4. Basile, C.; Combe, C.; Pizzarelli, F.; Covic, A.; Davenport, A.; Kanbay, M.; Kirmizis, D.; Schneditz, D.; van der Sande, F.; Mitra, S. Recommendations for the prevention, mitigation and containment of the emerging SARS-CoV-2 (COVID-19) pandemic in haemodialysis centres. *Nephrol. Dial. Transplant.* **2020**, *35*, 737–741. [CrossRef] [PubMed]

5. Sethuraman, N.; Jeremiah, S.S.; Ryo, A. Interpreting Diagnostic Tests for SARS-CoV-2. *JAMA* **2020**, *323*, 2249–2251. [CrossRef] [PubMed]

6. Long, C.; Xu, H.; Shen, Q.; Zhang, X.; Fan, B.; Wang, C.; Zeng, B.; Li, Z.; Li, X.; Li, H. Diagnosis of the Coronavirus disease (COVID-19): rRT-PCR or CT? *Eur. J. Radiol.* **2020**, *126*, 108961. [CrossRef]

7. Bataille, S.; Pedinielli, N.; Bergougnioux, J.-P. Could ferritin help the screening for COVID-19 in hemodialysis patients? *Kidney Int.* **2020**, *98*, 235–236. [CrossRef] [PubMed]

8. Thomas, A.; Silver, S.A.; Perl, J.; Freeman, M.; Slater, J.J.; Nash, D.M.; Vinegar, M.; McArthur, E.; Garg, A.X.; Harel, Z.; et al. The Frequency of Routine Blood Sampling and Patient Outcomes Among Maintenance Hemodialysis Recipients. *Am. J. Kidney Dis.* **2020**, *75*, 471–479. [CrossRef] [PubMed]

9. Lano, G.; Braconnier, A.; Bataille, S.; Cavaille, G.; Moussi-Frances, J.; Gondouin, B.; Bindi, P.; Nakhla, M.; Mansour, J.; Halin, P.; et al. Risk factors for severity of COVID-19 in chronic dialysis patients from a multicentre French cohort. *Clin. Kidney J.* **2020**, *13*, 878–888. [CrossRef]

10. Corman, V.M.; Landt, O.; Kaiser, M.; Molenkamp, R.; Meijer, A.; Chu, D.K.W.; Bleicker, T.; Brünink, S.; Schneider, J.; Schmidt, M.L.; et al. Detection of 2019 novel coronavirus (2019-nCoV) by real-time RT-PCR. *Eurosurveillance* **2020**, *25*, 2000045. [CrossRef] [PubMed]

11. Kita, H. Eosinophils: Multifaceted biological properties and roles in health and disease: Immunobiology of eosinophils. *Immunol. Rev.* **2011**, *242*, 161–177. [CrossRef] [PubMed]

12. Li, Q.; Ding, X.; Xia, G.; Chen, H.G.; Chen, F.; Geng, Z.; Xu, L.; Lei, S.; Pan, A.; Wang, L.; et al. Eosinopenia and elevated C-reactive protein facilitate triage of COVID-19 patients in fever clinic: A retrospective case-control study. *EClinicalMedicine* **2020**, *23*, 100375. [CrossRef] [PubMed]

13. Guan, W.-J.; Ni, Z.-Y.; Hu, Y.; Liang, W.-H.; Ou, C.-Q.; He, J.-X.; Liu, L.; Shan, H.; Lei, C.-L.; Hui, D.S.C.; et al. Clinical Characteristics of Coronavirus Disease 2019 in China. *N. Engl. J. Med.* **2020**, *382*, 1708–1720. [CrossRef]

14. Qarni, B.; Osman, M.A.; Levin, A.; Feehally, J.; Harris, D.; Jindal, K.; Olanrewaju, T.O.; Samimi, A.; Olah, M.E.; Braam, B.; et al. Kidney care in low- and middle-income countries. *Clin. Nephrol.* **2019**, *93*, 21–30. [CrossRef]

15. El Nahas, A.M.; Bello, A.K. Chronic kidney disease: The global challenge. *Lancet* **2005**, *365*, 331–340. [CrossRef]

MDPI

St. Alban-Anlage 66

4052 Basel

Switzerland

Tel. +41 61 683 77 34

Fax +41 61 302 89 18

www.mdpi.com

Journal of Clinical Medicine Editorial Office

E-mail: jcm@mdpi.com

www.mdpi.com/journal/jcm